THE PURSUIT OF SIMPLICITY

EDWARD TELLER

PEPPERDINE UNIVERSITY PRESS
Malibu, California

Copyright © 1980 by Edward Teller. All rights reserved. No part of this book may be reproduced in any form or by any means without permission in writing from the publisher, Pepperdine University Press, Malibu, California 90265. Printed in the United States of America.

ISBN 0-932612-02-4 (cloth edition), 0-932612-03-2 (paperback edition). Library of Congress Number 80-82499.

Produced for Pepperdine University Press by Guild of Tutors Press, International College, Los Angeles. Designed by Paul O. Proehl. Typesetting in Garamond by Kae McElwain.

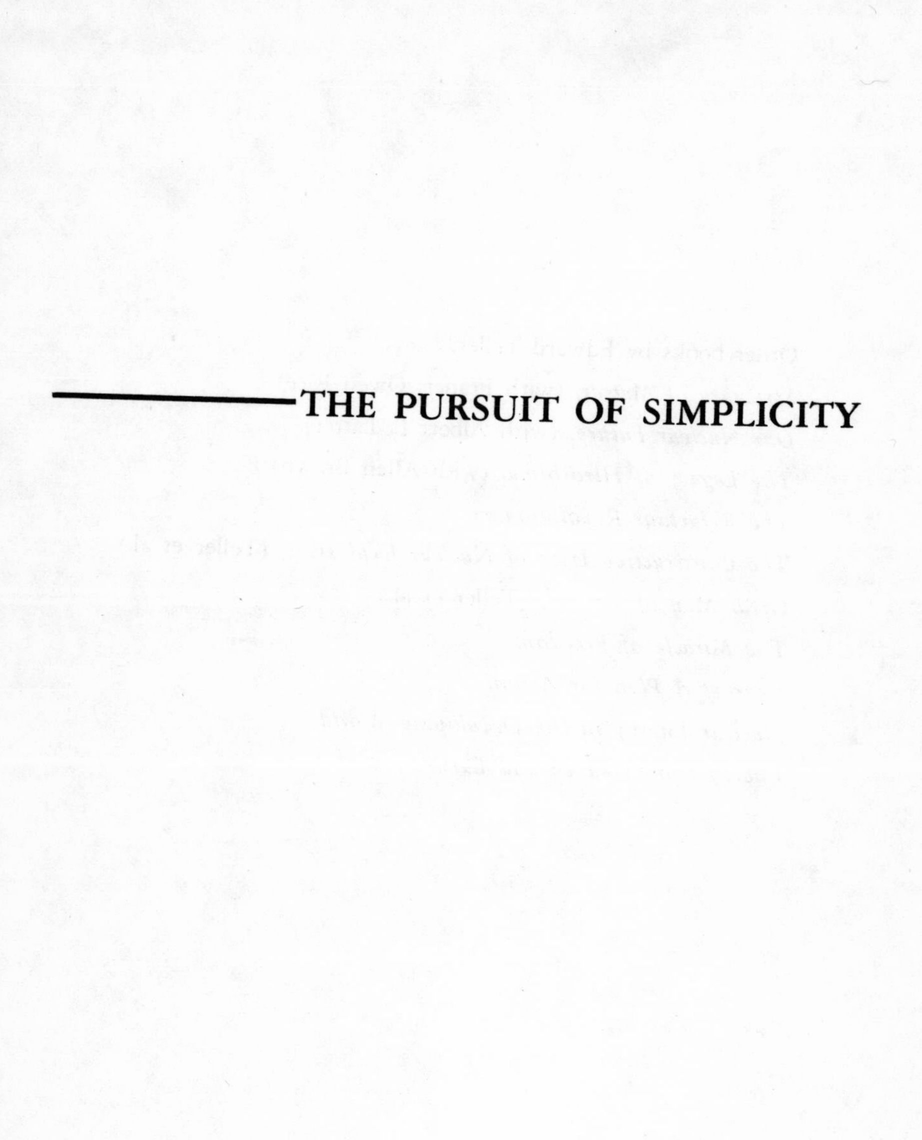

THE PURSUIT OF SIMPLICITY

Other books by Edward Teller:

Structure of Matter, (with Francis Owen Rice)

Our Nuclear Future, (with Albert L. Latter)

The Legacy of Hiroshima, (with Allen Brown)

The Reluctant Revolutionary

The Constructive Uses of Nuclear Explosives, (Teller et al.)

Great Men of Physics, (Teller et al.)

The Miracle of Freedom

Energy: A Plan for Action

Nuclear Energy in the Developing World

Energy from Heaven and Earth

This book is dedicated to
ARTHUR SPITZER
a true friend
of great and constructive
powers of imagination.

Table of Contents

Acknowledgments

This book began as a lecture series given at Pepperdine University in the spring of 1978. In the intervening period, Robert Budwine has been of invaluable help with his suggestions and work on turning a number of separate lectures into a coherent book. The changes have by no means been minor. I would also like to thank Keith Berwick for badly needed constructive criticism, Katrina Mommsen for her assistance in tracking down my Goethe quotes, Mark Shoolery for his assistance with the illustrations in Chapter One, and finally I want to thank my (merciless) editor, Judith Shoolery, for bringing about what I hardly hoped could be accomplished: a finished product.

Edward Teller
Hoover Institution
Stanford, California

Foreword

During the academic year 1978-79 the students and faculty of Pepperdine University were privileged to experience this book in the form of five lectures given under the auspices of the Arthur Spitzer Chair of Energy and Management. Hitherto the Spitzer Chair had focused its efforts on the organization of large-scale conferences to address the critical problems of energy and how they might be resolved. Now Dr. Teller was undertaking to put those problems in context of scientific development generally.

It is difficult to convey the special quality of those five lectures, the intellectual excitement they generated, the sense of the audience that they were participating in important events. Here was a master teacher, one of the world's great physicists, leading us through the labyrinthine passages of modern science with grace and wit and insight, and always with a profound appreciation of the human consequences of scientific endeavor. In the clarity of his vision Dr. Teller exemplified his major theme: that science is in essence the pursuit of simplicity.

Here then is a completed version of those lectures. *The Pursuit of Simplicity* is an inspiring book which reminds us of the central role that science plays in improving the human condition. It is a pleasure to share it with a wider audience.

Keith Berwick
Executive Director
Arthur Spitzer Chair
of Energy and Management

Introduction

THE LABYRINTH OF SIMPLICITY

To call a book on the development and applications of science *The Pursuit of Simplicity* may seem paradoxical. Science and technology are more often viewed as the source of complication in the world than as simplifying factors. Large numbers of educated people, even learned people, in the western world are gripped by a malaise of despair at a world that seems too complicated to understand, much less affect. They wish that the world were more simple, more as it used to be. Their pursuit of simplicity begins with wanting to discard science and technology. Rejecting science and technological progress, even if it were possible, would certainly not solve people's problems.

For most of the history of the United States, technological progress has been recognized as man's true ally. In the midst of a disastrous war more than 100 years ago, a wise Congress passed the Land Grant Act of 1862. The purpose, to make agricultural and mechanical arts available to young people, became the basis of American technological preeminence. It led to the remarkable circumstance that in a world hard pressed to feed its peoples, the United States accounts for 41 percent of the wheat and 71 percent of the corn available for sale on the world market.

In the last two decades there has been a substantial change of opinion about progress. People agree that the third world nations, the developing nations, suffer from poverty because they lack technology. It is agreed that providing technological training and assistance is one of the best means of offering long range protection against starvation, disease, and all the curses of marginal existence. Yet, there seems to be no agreement on calling technological progress good.

About the time that the Middle Ages were expiring, the legend of Dr. Faustus gained popularity in Europe. Faustus, an alchemist, was supposed to have superhuman powers because he had sold his soul to the Devil. A great many people considered the first stirring of the Industrial Revolution to be the Devil's work. The people today who want to enlighten everyone on the evils of progress may not realize that they are intellectual descendants of a medieval movement.

The remarkable concept of progress has many ramifications. In this book I will consider only one: the reputed role of progress in creating increased complication. I also hope to share an effective way to pursue simplicity, a course of action which men and women have developed over a long portion of human history. Its rewards are a consistency and predictability to the world and an opportunity to decide more effectively the course best suited to human needs. I am, of course, speaking of the use of science.

A First Paradox

Science has developed at an incredible rate in the last century. If science leads to simplicity, why is there now so much despair over the world's complexity?

Science has played a part in creating this malaise, but its role differs from what most people imagine. Science has not only developed new ideas at an incredible rate, but it has developed ideas that themselves seem incredible. This accumulation of unassimilated information is surely part of the problem. It has helped to create an "indigestion" which has the symptoms of anxiety, despair, and, occasionally, even hysterical fear.

There is more research today than ever before. Quite apart from the population explosion, there is an even bigger explosion of specialization. Solid state physicists work most of the time on electronics. Plasma physicists are concerned with the substance of which stars are made, most particularly with using this substance to release energy. Elementary particle physicists investigate exceedingly short-lived objects which are neither elementary nor particles, and which have never actually been seen. Turning from physics to psychology, one finds so great a multiplicity of chasms as almost to defy description on a single page of a book.

It is hard to avoid recalling the time of Babylon when the builders of the Tower lost the ability to talk with each other. Volumes and volumes are published today where even the single words can no longer be understood except by specialists in specific fields. In addition, many words have been adopted for special purposes. What the word *group* means in everyday language, the reader will know. When a psychologist uses the word, he is reminded of the differences, for example, among a family, club and mob. What *group* means to a mathematician is even further removed. The mathematician "multiplies" members of a group and obtains new members of the same group.[1] Peculiarly enough, the definition of the mathematician is the sharpest and is remarkably useful, provided, of course, that one is a mathematician.

The inability to exchange ideas is not only disturbing, it is dangerous. Twenty years ago, C. P. Snow published a book, *The Two Cultures,* about the growing ideological gulf between people of science and people of letters and arts. He was concerned that people who need to make decisions jointly had ceased talking about the same realities. Not only were the terms different, but even the major ideas were not mutually recognized. A democratic society must have a common understanding of ideas and their consequences in order to work together for the future. Yet the situation today does not seem to have improved.

A Spool of Thread for the Labyrinth

There can be no question but that modern research heaps observation upon observation, pours new facts into an ocean already brimful of details that no one can remember and seemingly no one can hope to digest. The purpose of science is to find simplicity and coherence in the billowing mass of material. In this sense, science (as distinguished from raw research) serves as a guide. The details may remain in reference libraries or on the memory discs of electronic computers.

What is science? My definition includes a qualification not commonly found in dictionaries. I would say that science is a set of relevant, consistent statements of general validity that also contain an element of surprise. Science is almost always surprising, startling, even an affront to conventional perceptions. This element of surprise has given rise to difficulties and excitement among scientists. It has also managed to trouble and confuse most of the non-scientists.

[1] In this kind of multiplication, a times b is not equal to b times a.

Most people don't think of science as surprising, but they will generally agree that when first encountered a scientific statement is not obvious. It will seem obvious only after it is considered in relation to many other facts and explanations. The characteristic novelty in science has always produced an initial feeling of disbelief, sometimes even of consternation. The feeling that the world is too complicated to understand is not uncommon in human history. Incredible information has always been particularly frightening when it deals with ideas that cannot be verified or discovered in a simple direct way.

Yet one of the most general and basic drives in human beings is the desire to understand. The question *why* is the first sign of human nature in children. It is a question that needs a careful response. In my opinion, the best response is the one that will cause the person to repeat the questioning process. For most of man's existence, the response to human curiosity has been the invention of an incredible variety of myths and fables. They are charming, fascinating and often profound. Yet all of them lack consistency and simplicity. Science is a fable which has been made consistent.

Simplicity, for me, is best characterized in a story from the art traditionally the favorite of mathematicians and scientists: music. When Mozart was fourteen years old, he listened to a secret mass in Rome, Allegri's *Miserere*. The composition had been guarded as a mystery; the singers were not allowed to transcribe it on pain of excommunication. Mozart heard it only once. He was then able to reproduce the entire score.

Let no one think that this was exclusively a feat of prodigious memory. The mass was a piece of art and, as such, had threads of simplicity. The structure is the essence of art. The child who was to become one of the world's greatest composers may not have been able to remember the details of this complicated work, but he could identify the threads, remember them and reinvent the details having listened once with consummate attention. These threads are not easily discovered in music or in science. Indeed, they usually can be discerned only with effort and training. Yet the underlying simplicity exists and once found makes new and more powerful relationships possible.

The Ultimate Dream: Simplification Completed

There are many questions about the world which are unsolved. Not surprisingly, they are the biggest questions of all for most people, including me. What is life? What is consciousness? What are human beings? These questions may never have answers. Someday they may be understood as having been the wrong questions to pose; they may need to be asked in terms that today would appear peculiar to everyone. I offer an interim answer to one of these great enigmas: life is a little matter endowed with an enormous purposeful complexity.

My answer has much in common with artless myths. Like them, it postulates what it pretends to explain. In early fantasies about creation, the world was explained in terms of gods modeled after humans or even after the animals the gods were going to create. In my definition of life, a little present day knowledge is included. I use the word *matter* in good conscience because I shall discuss matter in the third chapter. In talking about *complication,* I am on less certain ground. The adjective *purposeful*—now that word is truly dangerous. What purpose? Whose purpose? At this point, I am completely beyond my depth. What I want to point out is that to be alive implies complication which is handed on from generation to generation.

Simple observation, medically elaborate observation, even more, observation in molecular biology, shows that complication is a fundamental property of life. Simplicity is of interest only as an opposite to unnecessary complication; taken by itself, it is nothing more than monotony. The scientific point of view, in its orientation and consistency, differs from myths, legends and poetry and a host of magnificent human achievements. Science introduces consistency and simplicity into a world that without them appears confused, random and even whimsical.

However, the surprises of science are not easy to accept even when one devotes real attention to these ideas. Many years ago as a high school student I wrote a statement in my Hungarian diary that reflected despair over my confusion:

> What is called understanding
> is often no more than a state
> where one has become familiar
> with what one does not understand.

In the course of time, I came to realize that this view is not one of despair but of hope. It is possible to get used to a set of facts by thoroughly exploring their interrelationships. These strange facts can become so familiar that one no longer feels lost among them. When one can use them to predict, then one can say, "I understand," and understanding is one of the great human comforts.

Domesticated Paradoxes

A recent "Golden Age" in science has given mankind amazing new concepts and abilities. Unfortunately, this new gold has quite limited currency except among a small group of connoisseurs. I hope in this book to present some of these scientific developments in such a way that even the person who knows little mathematics will be able to grasp their essence. I shall cover five very different topics. I would like to convince the reader, at least partially, that simplicity is present in the science of the modern world and has, in fact, increased. Moreover, I should like to persuade the reader that science is a group of findings and methods that everyone can use with some success to simplify, to "domesticate," the world's paradoxes.

The first chapter concerns the earth and the solar system. The idea that the earth is a sphere moving in space was not always easy for people to accept. (The Devil was seen as inspiring this idea, too.) However, everyone in most of the world today has seen a globe and become familiar with the idea of an earth traveling around the sun. People have somehow developed what can be called understanding. Yet, having observed the "flat" earth, one should retain a feeling of wonder that the people at the antipodes are walking upside down. Few ideas are ingrained as deeply in human thinking as *above* and *below*. These ideas come from direct experience and are not consistent in an obvious way with the earth being a ball.

There is, of course, much more to the current concept of the earth and solar system. The Copernican revolution eventually produced an important insight transcending the simpler questions of planetary motion. The development of astronomy led to a monumental issue, to the concept that laws of nature exist, laws that hold true regardless of whether one considers the earth or the heavens. This history and a glimpse of the partial answer it offers to the questions of the nature and beginning of the universe constitute the first chapter.

The state of motion is indistinguishable from the state of rest! Time assumes different values for different observers! Matter warps space! These statements are part of one of the most famous and least understood developments of modern science, the subject of the second chapter. Why are these ideas still so obscure for most people three-quarters of a century after they were first presented? The reason is not mere mathematical complexity.

Furthermore, an essential part of the theory—the importance of the perspective of an observer—is an idea most harmonious with an individualistic philosophy. Translated from physics to politics, the theory would say that every observer considers his or her perspective moderate (that is, in the central position). Others' views are slanted, even extreme. The amount of bias perceived in another's viewpoint is dependent on the (political) distance between observers and the speed and direction of movement (change of opinion) of each. In politics and physics, right and left are relative.

Relativity itself is not complicated. However, it is very surprising. Some of the tenets are actually startling when one tries to understand them on the basis of common experience and intuition. The truth and systematic simplicity of relativity manifest themselves only after one manages to recover from the first amazement. The man who laid the foundations of the theory was, of course, Albert Einstein, and the main products of his work form the content of the second chapter.

The next big step in understanding is into the world of atoms. The theory of atoms demands in a most profound way that familiar visualizations be relinquished. Ironically, Einstein contributed a great deal to the early development of atomic science, but he could never accept its surprising conclusions. These conclusions are most peculiar when judged by common experiences. They call into question the basic law of causality, the straightforward connection between cause and effect.

Classically, science was dedicated to the concept of events linked by an unbroken chain of consequences of prior events. Edward Fitz-Gerald, an English poet of the 19th century, was inspired by the work of a poet-scientist who lived more than seven centuries earlier in Persia. In the second edition of *The Rubaiyat of Omar Khayyam,* FitzGerald wrote:

> For let Philosopher and Doctor preach
> Of what they will, and what they will not—each
> Is but one Link in an eternal Chain
> That none can slip, nor break, nor over-reach.[2]

[2] The popular opinion that Omar Khayyam had a clear idea of determinism appears to be erroneous. One is reminded of Goethe's dictum:

> Was Ihr den Geist der Zeiten heisst
> Das ist im Grund der Herren eigner Geist
> In dem die Zeiten sich bespiegeln. . .

which I would translate (with less license than FitzGerald used):

> What you have called the Spirit of an Age
> Is just the spirit of a later sage
> For whom the past is but a mirror.

If science denies this eternal chain, if there is no such rigid connection between cause and effect, must people give up hope of ever recognizing order in nature? No! Every gambler knows that chance also has laws. The scientists whose entire mental training and outlook are contrary to chance may find it harder to accept this concept than the layman, but once the step is taken, here again is a source of novel simplicity.

In the study of the very small constituents of matter, the means of observation are the same size as the observed object. In this case, the links between cause and effect become more intricate. Here it is necessary to discuss the "science of uncertainty." This paradoxical concept was propelled into physics by Niels Bohr, a man dedicated to paradoxes. The third chapter takes the reader about as far as the solid recognition of the physical world has developed.

Science attempts to find logic and simplicity in nature. Mathematics attempts to establish order and simplicity in human thought. This simplicity is based on elementary ideas such as numbers, triangles or the rules of logic. Doesn't such simplicity introduce both boredom and overwhelming complication? The actual work of the mathematician appears exceedingly complicated to most everyone. However, the cumulative and complicated application of well defined ideas found in mathematics also produces threads of simplicity.

The increased understanding of the intricate nature of matter (and some help from mathematicians) made it possible to construct "thinking machines." These machines can perform functions of the human thinking process with greater speed and considerable reliability—provided that humans can spell out what that thinking process is.[3] Computers not only relieve mankind of repetitious chores and lead to new applications of mathematics, but they also open up entirely new vistas. The question of what constitutes humor is deceptively easy, as are the questions of how one balances on a bicycle or translates a language. Through the use of computers, it may be possible (in the not too distant future) to gain a little more understanding of these processes. When questions like these can be answered, there may be some hope of fulfilling Plato's ancient admonition: *know yourself.*

Economists, historians, and political scientists are using this new tool, the computer. I have little doubt that they will produce many new complications, have to turn back from many blind alleys before they have any chance of finding the simple statements that one may

[3] This is a rather severe restriction since, in general, people are rather thoughtless about their thinking process.

call truths within the most complex of all systems, those composed of aggregates of human beings. Ultimately though, technology may help simplify even this most massive complexity. Undeniably technology is becoming more complex. Yet the most incredibly complex machines should be less disturbing than the knowledge that people themselves may be hopelessly complex. The mathematics performed by man and machine and some consequences of the man-machine relationship are the subject of the fourth chapter.

Semi-Domesticated Paradoxes

In discussing thinking machines, I have only touched on one way in which the application of science, technology, has affected daily life. The fifth chapter approaches a problem of a completely different nature. Science and technology have changed everyone's mode of living. In the next few decades this scientific and technological revolution will most likely spread over the face of the earth. The result of this development is of the most obvious interest to all the world's inhabitants.

Science has made most people aware of the reality of the invisible, of things and forces and orders that lie beyond the ability of one's direct senses to perceive, such as the atomic nucleus. Many consider the consequences of this knowledge not only with uncomprehending amazement but also with an almost physical fear of the unknown. As in ancient myths, some people predict the end of the human race. Even some scientists, whose predictions should be more temperate and limited, are not immune to the sentiments which permeate society.

When I try to look into the future, I am again brought back to ideas and questions connected with simplicity. The future will be influenced, hopefully, by all the billions of people on the planet. Unfortunately, the international paradoxes are, at best, only semi-domesticated. Without some simple ideas concerning the future, mankind cannot collectively have a reasonable goal. There are some simple statements in a familiar document, truths so simple that in 1776 they were considered self-evident. Those truths made possible the building, rather than the evolution, of a great nation.

Today, in an interconnected, interdependent world, it seems equally self-evident that the importance of technology and science to mankind, both in the past and in the future, must be acknowledged. Each individual should hold evident that these human activities can work for the betterment of humanity. Each should hold evident that the power which people are rapidly acquiring over nature and potentially over one another must be subject to mutual understanding and agreements between nations.

To me, these statements are simple and self-evident. Whether there can be agreement on them is perhaps the most important question to be decided in the next few decades. A different and contrary claim is that the future should be planned by logical principles and thereafter enforced upon everyone. What agreements can be reached between a society built on individuality and a social order enforced upon individuals? Are these two approaches irreconcilable? Will a decision be made without the use of the age-old arbiter, violence? These are some of the subjects considered in the last chapter.

In none of this discussion do I mean to imply that the world or life or the future can be simple. A more modest and realistic claim is made: to pursue simplicity in life, in the world, for the future, is a most valuable enterprise.

Chapter One

THE LAWS ON EARTH AND IN THE HEAVENS

Science is (or ought to be) simple, but its essence is to be surprising. Contemporary science, rightly understood, is both simple and surprising, but the beginnings of science were quite different. As perceived today, science is basically a Greek invention and began when "proofs" entered man's conscious thinking. A proof in science does more than eliminate doubt. It eliminates inconsistencies and provides the underlying logical basis of the statement. Knowledge, of course, is much older.

Science Begins With A Proof

Babylonians were using the theorem credited to Pythagoras a thousand years before this philosopher-mathematician was born. However, Pythagoras found the proof, and today that discovery is seen as the beginning of geometry. Proofs were common in Greek mathematics, but they did not characterize the rest of Greek science (which is more aptly called philosophy). Much early science was the opposite of surprising and simple—it was obvious and complicated.

Aristotle, a leader of Greek learning with a passion for encyclopedic knowledge, described the earth as unmoving. Even so, the Greeks soon discovered that the earth was not what it seemed. They found that the earth was a sphere, and that was a big surprise. The Greeks also knew quite accurately just how big our sphere is.[1]

The Greeks noticed that the slant of the sun's rays changed as one moved from north to south. The difference of the angle at two points on earth, one directly south of the other, could be easily observed at noon during the summer solstice (when the sun appears to come as far north as it ever does). If one knows this angle and the distance between the points of measurement, one can determine the circumference of the earth. Eratosthenes is credited with originating this principle and first applying it. To measure the angle, he used a perpendicular stick in Alexandria and a well with vertical walls in Syrene (modern Aswan). (See Figure 1.1.) The stick in Alexandria cast a shadow that indicated the sun was 7° 12′ from the vertical. At the same time, the sun was shining directly down the well at Syrene. The distance between the well and the stick was 493 miles. Since 7° 12′ is 1/50 of 360°, the circumference of the earth was calculated to be 50 x 493 miles, or 24,661 miles. This is only about 150 miles less than today's best measurement.

The Greeks considered the earth the fixed center of the universe; the moon, planets, sun and stars revolved around it. They, like other ancient peoples, observed that an impressive star-studded sphere rotates in a uniform manner around the earth. Circular motion was therefore introduced as the basic movement of heavenly bodies. This apparently eternal firmament, undeviating in rotation, was known as the *primum mobile* (loosely, the original mover) and was also described as the *seventh heaven*, the residence of the Gods. But there are objects between the seventh heaven and the earth. The planets are "wanderers" among the other stars. The sun and, even more, the moon change their positions on the background of the myriads of apparently lesser stars.

Because circular motion with constant speed was accepted as fundamental, the motion of objects in the intermediate heavens had to be explained as composed of uniform rotations. The planetary motion that had actually been observed is not simply circular. The problem was solved by introducing circular motions whose centers were moving on circles. Such an orbit is called an *epicycle*, and very

[1] The measurement, made a little after Aristotle's time, was effectively lost, though the concept of a spherical earth survived. The size of the earth had to be rediscovered by western cultures. Columbus greatly underestimated the size of the globe when he set out on his great adventure. Had he known the distance between Spain and China, no doubt he would have stayed at home. His older brother, who worked in Henry the Navigator's great research and development project at Sagres (probably the first of its kind in history), most likely knew the answer.

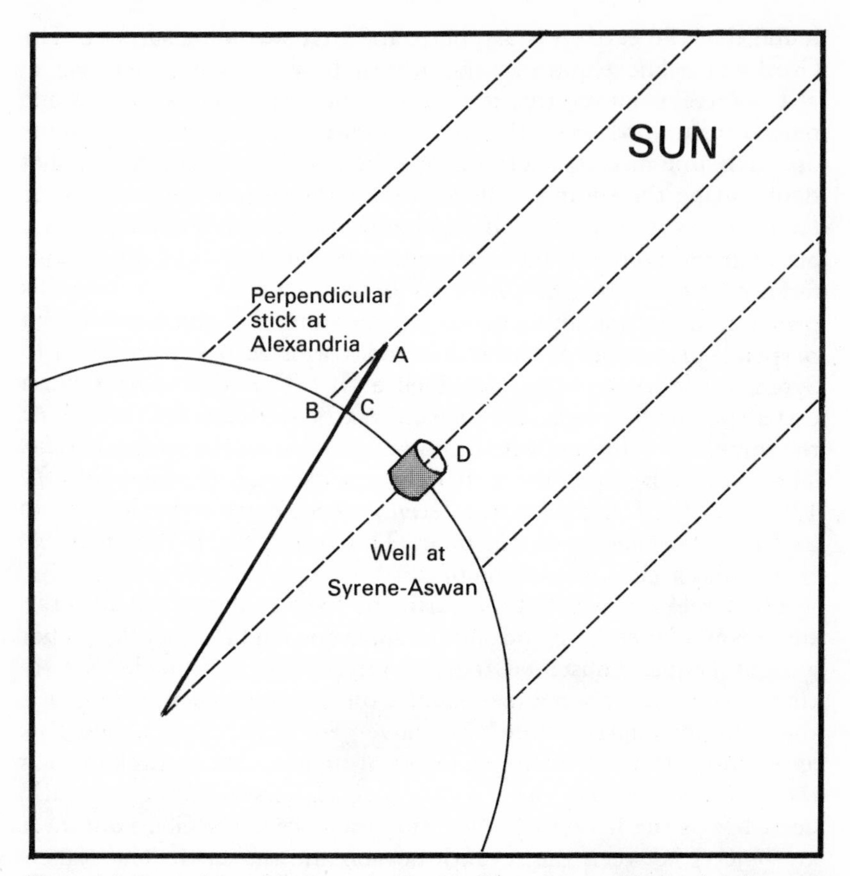

Figure 1.1. Eratosthenes Measures the Earth's Circumference. Notice that the angle BAC is the same as the angle between the solid and the broken line at the center of the earth. The distance BC is, in fact, negligible.

complex epicycles were needed. For example, the planet Mars was said to be moving on a circle whose center was moving on a circle, whose center was moving on a circle. . . repeated a total of five times. It is not easy to consider Aristotelian science simple.

The final version of the epicycle theory was put together in the 2nd century A.D. in Alexandria by Ptolemy. These ideas would have been lost except that the Arabs who conquered the Hellenistic world centuries later were interested in Ptolemy's description of these wheels within wheels. They accepted and reproduced it, and thus Ptolemy's work, *Almagest* (the Arabic prefix *al*, combined with a misspelling of *majestic*), has come down to us.

The first truly surprising and basically simple proposal to escape from this jungle of complication was made at a much earlier time. In the 3rd century B.C., again in Alexandria, a Greek, Aristarchos of Samos,[2] proposed that the earth is not at rest but revolves about its axis and moves on a circle around the sun. The moon, Aristarchos explained, rotates around the earth, but the planets, like the earth, orbit the great conspicuous central body of the sun. For Aristarchos, the remarkable rotation of the seventh heaven was appearance rather than reality.

Aristarchos' actual writings on this topic have been lost. His idea was considered so absurd that it was almost completely ignored. His heliocentric concept has survived only as a brief comment in Archimedes' writings. This great mathematician had the good judgment to recognize the idea as remarkable enough at least to mention. Aristarchos' fate here is in contrast to Galileo's: one astronomer attracted too little attention, the other too much, for espousing the same view.

No one knows how Aristarchos arrived at his radical idea, but it may have been connected with another important contribution that he made. Aristarchos was more than an idea man—he was an excellent practical astronomer. He devised a method to measure the distance of the sun from the earth, and his work on this problem is known in reasonably complete form.

Aristarchos' contemporaries knew the shape and size of the earth. They had also determined the distance of the earth from the moon by means of earth-bound parallax. Parallax is the phenomenon used in everyday life to estimate the distance of an object with the help of binocular vision. The technique is easily demonstrated by looking at one's extended index finger relative to background objects. The apparent motion of the finger when one looks first with one eye and then with the other gives an approximation of the distance of the index finger from one's eye. The greater the apparent motion of the finger against the background, the closer the finger is to one's eye. In the case of this measurement, the moon would represent the index finger and the stars the background objects. The closer the moon, the more it would appear to move on the background of the stars when viewed from different places on the earth. It is common knowledge today that the moon is little more than 200,000 miles away. The Greeks knew this too, though not as accurately.

[2] Aristarchos' name (spelled here in conformity with its Greek derivation rather than in the Latin form more commonly used) comes from two root words: *aristos*, meaning *best*, and *archos*, meaning *origin*. Thus Aristarchos' name, like his idea, may be remembered as a best beginning.

One could try to use the same trick to determine the distance to the sun. Unfortunately, the sun is about 400 times as far away as the moon, and the parallax from two points on earth is too small to be observed by any of the old, simple methods. In addition, it is practically impossible to see the background stars and the sun at the same time.

Because of these problems, Aristarchos devised a very ingenious method that did not depend on parallax. An observer on earth would identify the moon as being in its first or third quarter when the moon appears to be half-illuminated and half-dark. Between the first and third quarter, the moon moves to the approximately opposite side of its orbit of the earth, a considerable distance in space. If the sun were infinitely far away, the sunlight would always strike the moon at the same angle, regardless of the moon's position. In this case, the first and last quarters of the moon would occur at exactly equal lengths of time from the new and full moon. (See Figure 1.2.a.) Or, one could say, arc AB = arc BC = arc CD = arc AD because the rays of the sun coming from an infinite distance would be perfectly parallel.

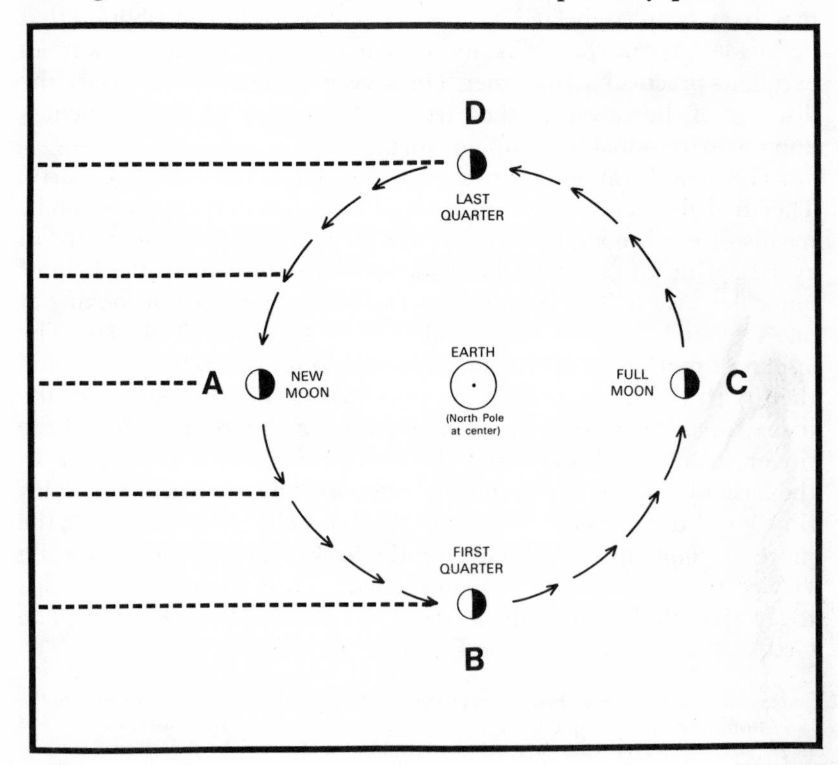

Figure 1.2.a. The Phases of the Moon If the Sun Were Infinitely Far Away.

Light rays coming from a source a finite distance away would have a slight angle. This angle would cause the moment when the sun's rays divided the moon exactly in half to occur slightly closer to the new moon than to the full moon. (See Figure 1.2.b.) The size of the angle of the sunlight could be determined by measuring the time differences between the phases. Using the known distance of the moon from earth and the angle of the sun's light, Aristarchos derived the distance between the earth and the sun.

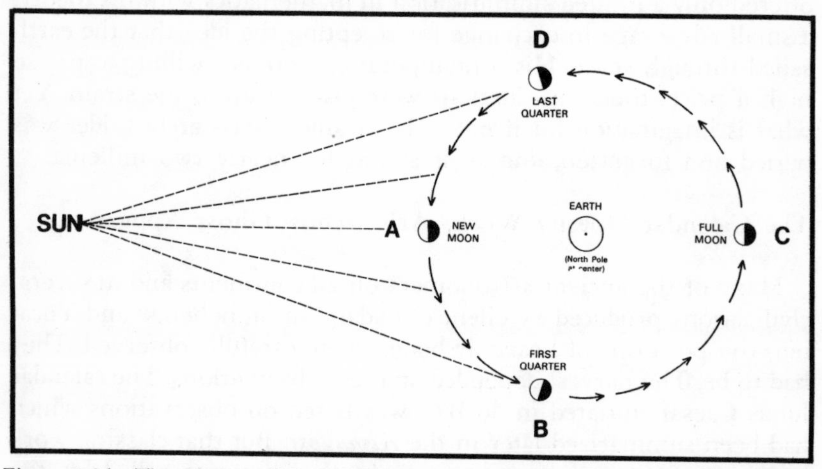

Figure 1.2.b. The Phases of the Moon If the Sun Were a Finite Distance Away.

The principle was right; the measurement was quite inaccurate. It is very difficult to determine when the moon is precisely half-full because the uneven surface of the moon distorts the shadow. Aristarchos made a very slight mistake, and in this case, even a slight mistake can lead to a big error. He calculated that the sun was six times closer than it actually is. This error was not corrected by astronomers until after the days of Galileo. People believed Aristarchos' measurements for almost 2,000 years; his wonderful heliocentric system they forgot. (See Appendix I.)

Aristarchos' argument for his theory may have been no more than common sense. He obviously knew from his measurement that the sun is very much larger than the earth (even though he underestimated the distance). He could have argued that it is absurd to believe that the much bigger object should move around the smaller one: it would be a case of the tail wagging the dog. It was much more reasonable that the small earth should move around the big sun. That may well have been all there was to it, but what Aristarchos proposed was surprising, unconventional and totally unacceptable to his contemporaries.

Perhaps one can say something in defense of the conservative school which rejected the heliocentric view. If the earth and planets did move in perfect circles around the sun (and the moon similarly described a perfect circle around the earth), then Aristarchos' explanation would have fit the facts exactly. However, the orbits are not circles but ellipses, a fact unsuspected at that time. So Aristarchos offered only a limited simplification in mathematics with his theory, a small advantage in exchange for accepting the idea that the earth sailed through space. His contemporaries were not willing to pay so high a price; their imaginations were just not up to the strain. Yet what is imagination for if not to be strained? Aristarchos' idea was buried and forgotten, and so it stayed for nearly two millenia.

The Calendar Doesn't Work: Aristarchos' Ghost Appears

Many of the ancient astronomers on all continents and in several civilizations produced excellent calendars. In Stonehenge and Yucatan, the positions of heavenly bodies were carefully observed. They had to be. The harvest depended on these observations. The calendar Julius Caesar initiated in 46 B.C. was based on observations which had been summarized later in the *Almagest.* But that classical work had been corrupted in recopying. As the centuries rolled by, the Julian calendar proved good but less than perfect. Astronomical observations and seasons appeared to be slightly out of joint. A job of renovation was needed.

The major problem was that the calendar year exceeded the true solar year by 11 minutes and 14 seconds. By 1582, this inaccuracy required that ten days be dropped, so the day after October 4, 1582, became October 15, 1582. This marked the establishment of the Gregorian calendar that is used today.[3] The Julian calendar was preserved by the Greek Orthodox Church, however, and was used in Russia even in this century.

Prior to the changes made by Pope Gregory XIII, a council was convened to make recommendations for the calendar reform. A

[3] England was not on good terms with Rome in 1582 and so did not adopt Pope Gregory's calendar. In 1752, this country was obliged to drop eleven days when it finally made the reform. An independent peculiarity was that the English had celebrated New Year's Day on March 23, so that, for example, March 22, 1700, was followed by March 23, 1701. Making January first New Year's Day is therefore a relatively recent invention. One consequence of all these reforms was that George Washington had to change both the day and the year of his birth. I leave it to the reader as to what these changes were.

Polish canon, Nicolas Copernicus, was invited to attend but declined because he believed that the movements of the sun and the moon were not sufficiently well determined to make a really accurate calendar. However, Copernicus promised to study the question diligently and made a literature search of the classical writings. In the process, he came across a statement which Archimedes had written in a little treatise called *The Sand Reckoner*:[4] "For he (Aristarchos of Samos) supposed that the fixed stars and the sun are immovable, but that the earth is carried round the sun in a circle. . ."

Copernicus appears to have been rather conservative, even a plodding sort of man, but he noticed that the epicycles would be much simpler if one adopted Aristarchos' heliocentric view. His approach here was quite different from that of the Greek philosophers for whom systems of thought were all important. For Aristotle, not explaining everything meant that one had explained nothing. Copernicus was anything but a philosopher of this sort. He was engaged in a meticulous enterprise to bring the *Almagest* and the stars and seasons into consonance. The resurrection of Aristarchos' idea meant less complicated epicycles. Heliocentrism, even with circular orbits, means that the main motion of heavenly bodies takes place on circles to which relatively minor corrections must be added. The most conspicuous motion is simply due to the rotation of the earth around its axis. Once these adjustments were made, the remaining calculations (even with the epicycles) became much easier.

So Copernicus assumed that celestial bodies were moving around the sun. What awakened Aristarchos' beautiful idea from its long slumber was that it made correcting the corrupted *Almagest* less difficult. It was simplifying the struggle with the real calendar problem, rather than the simplicity—inherent but not apparent—of the idea itself, that made the difference in the end.

While Copernicus was willing to use this idea, he did not want to stand up to the criticism that its espousal would bring. He published his ideas only as he was dying, and even then an apologetic foreword warned the reader:

> And if any causes are devised by the imagination, as indeed many are, they are not put forward to convince anyone that they are true, but merely to provide a correct basis for calculation. Now when from time to time there are offered for one and the same motion different hypotheses. . . the astronomer will accept above all others the one which is the easiest to grasp. The

[4] This book contains many imaginative and provocative ideas. Pythagoras may have been the first to introduce proofs in science, but Archimedes was the first to emphasize surprise.

> philosopher will perhaps seek the semblance of the truth. . . . So far as hypotheses are concerned, let no one expect anything certain from astronomy, which cannot furnish it, lest he accept as truth ideas conceived for another purpose, and depart from this study a greater fool than when he entered.

Truth appears here as a very modest maiden indeed. It will not be very long before she becomes bolder.

Copernicus' great work was published, at the urging of his students and co-workers, in 1543: *De Orbium Coelestium Revolutionibus (Of the Revolution of the Celestial Spheres).* It did not stir much excitement in the Catholic Church, partly because most Catholics did not read Latin. Luther and his followers (who advocated that everyone read the Bible) were far more vocal and critical. Copernicus' work was not banned by the Catholic Church until 1616, more than seventy years after it was published. It was the Protestants who first raised the cry of "heresy."

In 1546, Tycho de Brahe, the greatest experimental astronomer of this time, was born in Denmark. He had no telescope but worked with the aid of other relatively primitive but painstakingly well-constructed instruments such as the astrolabe. With the help of a fabulous observatory built on the island of Hveen (given him by the King of Denmark), he measured the positions of the stars, moon and planets with unprecedented accuracy.

De Brahe thought, if Copernicus is right, then as the earth moves on its enormous circle, one should be able to see the nearer stars move with respect to the more distant stars because of the parallax. He compared observations taken in the summer and winter as accurately as he could, but he found no movement. Therefore, he concluded that Copernicus' hypothesis must be wrong.

The parallax that de Brahe tried to measure is so small and hard to detect that it was not observed until 1838. So de Brahe—purely from his experimental point of view, uninfluenced by religion, prejudice or tradition—adopted his own model in which he assumed that the earth was in a fixed position. He had no quarrel with the earth rotating around its own axis, but he assumed that the sun and moon circled the earth and that all other planets were moving around the sun. This compromise was considered seriously by the Catholic Church long after de Brahe's death.

An Astrologer-Mathematician Finds the Ellipse

The effective beginning of the Copernican revolution came with a very unusual German, perhaps a little crazy but wonderful, the half-astronomer, half-astrologer, Johannes Kepler.[5] In order to understand his work, one should know that Kepler was completely dedicated to understanding the solar system as the handiwork of God. He sought employment with Tycho de Brahe because he wanted access to de Brahe's extremely extensive and precise observational data. Kepler's eyesight was extremely poor. The stars, the objects of his greatest interest, he could follow only through the eyes of others.

De Brahe was a rather irascible person, and when a new king was crowned in Denmark, the two had a falling out. Tycho threatened to leave once too often, and the king helped him pack. Kepler became de Brahe's assistant shortly after this, when de Brahe was in Prague, employed by the Hapsburg emperor, Rudolph II. Eighteen months later, de Brahe died. His papers should have gone to his legal heirs, but Kepler stole the papers! Legally, of course, Kepler was wrong, but the scientific results are such that they seem to justify the means. Kepler declared that he needed the papers—that he could do something with them. Indeed, he did. He spent his life on them, finding the precise mathematical description of the motions de Brahe observed.

Kepler started with the most difficult part of the problem because he believed that if he could solve this, he could solve the rest. By the same principle, Bohr should have started his work on the uranium atom instead of on hydrogen. Fortunately, Bohr did not. Kepler's approach is not the usual one in science, but in great scientific achievements, it is often the unusual approach that counts. Aristarchos almost succeeded in explaining all of the planetary motions. The important point now was to clarify the remaining small deviations. It was not so crazy to concentrate on the case where these deviations are most conspicuous.

By starting as he did, Kepler was able to see the inherent peculiarities of the motions of the solar system. Mars' orbit exhibits the most peculiar motion and so required the most complicated set of epicycles. Kepler put practically all his effort into trying to interpret it. His struggle is beautifully described in Arthur Koestler's book,

[5] Kepler truly was half-astrologer. He looked for magic formulas, for relationships between the positions of planets and events on earth. He earned part of his living by casting the horoscopes of Emperor Rudolph and General Wallenstein during the Thirty Years War.

The Sleepwalkers (on which some of the material in this chapter is based). Koestler titles the section on Kepler's work "The Watershed." As this suggests, Kepler stood at the point between mystical astrology and logical astronomy. It may have been this tension that drove him on in an illogical, passionate and successful way.

Reading Kepler, one has a sense of anything but order. Kepler describes a hypothesis in one chapter and includes pages and pages of detailed tables. He begins the next chapter:

> Who would have thought it possible? This hypothesis, which so closely agrees with the observed oppositions, is nevertheless false. . . .

In another instance, Kepler discusses an error:

> What happened to me confirms the old proverb: a bitch in a hurry produces blind pups. . . . When these ideas fell on me, I had already celebrated my new triumph. . . without being disturbed by the question. . . whether the figures tally or not. . . .

This is not only amusing reading, but in some ways is much more honest than the publications of contemporary science.

In his continuing attempt to refine his calculations, Kepler almost managed to explain the motion of Mars with the help of the modified Copernican epicycles. But there were some deviations from the observations in this design, deviations only twice as big as the errors in de Brahe's observations. Kepler had almost succeeded, but because he believed celestial revolutions were God's handiwork, they had to be perfect.

Now the obvious thing to do would have been to use one more circle in constructing the epicycles and eliminate the remaining discrepancy. But at that point Kepler threw out his many years of work. God wouldn't have constructed the universe in such an inelegant and complicated manner. Using another epicycle meant that the problem could have been solved in many ways. For God to have made His universe in a less than unique manner was intolerable. In any case, Kepler felt there had to be a simpler way, so out with the epicycles. Kepler spent many hard years on epicycles only to get rid of them.

He looked for curves that might be better suited, and he found what he needed. The Greeks had described other curves, one of which was the ellipse. (See **Figure** 1.3.) An ellipse is a circle viewed from the side. It appears flattened when compared with a circle. Through Kepler's work, the circle lost its privileged status as the perfect curve.

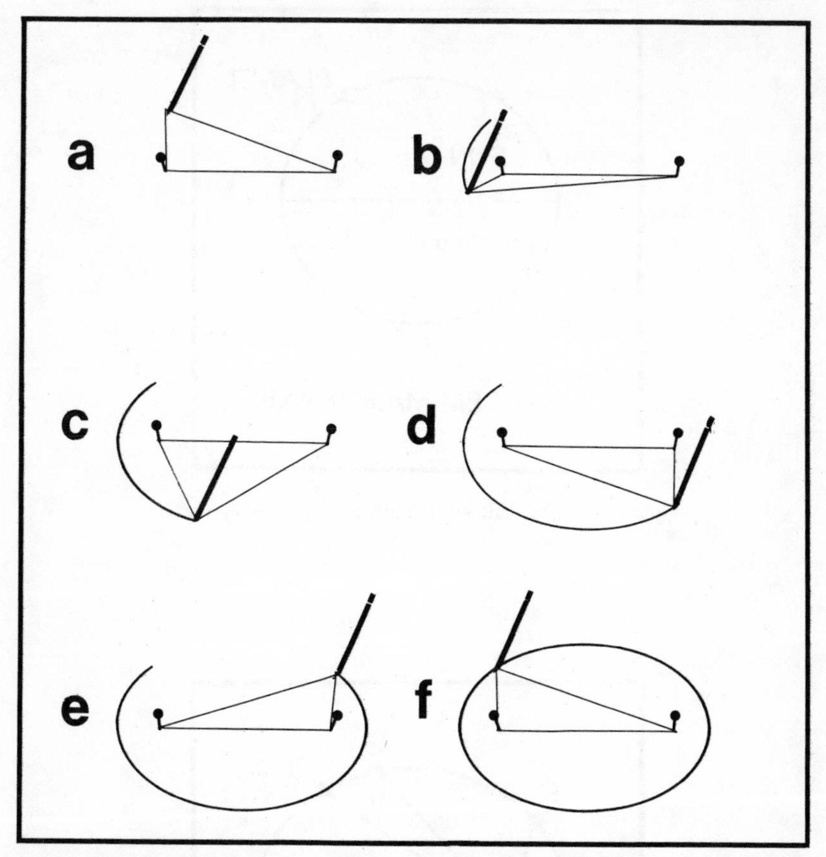

Figure 1.3. The Shape of An Ellipse. The regularity of an ellipse can be easily understood by drawing one. Place a loop of string around two thumbtacks and pull it taut with the tip of a pencil. Keeping the string taut, attempt to draw a curve. The result will be an ellipse. The two tacks are in the position of the foci of the ellipse.

It was obvious that the sphere of the stars was not moving at all; the earth was rotating. In addition, the curve on which the planets traveled embraced two foci rather than a single center. The sun was at one of the foci, and the planets and the earth moved on various elliptical curves. This is Kepler's First Law: every planet travels an elliptical orbit with the sun at one focus of the ellipse. (See Figure 1.4.)

Furthermore, the speed of the planets was not constant. Perhaps some other quantity would not vary. Kepler found that a line drawn from the sun to a planet will sweep out equal areas in equal times. This means that when the planet is far away from the sun, it moves

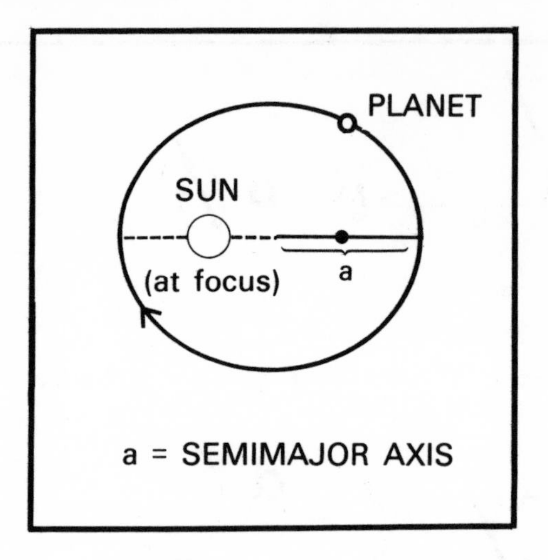

Figure 1.4. The Elliptical Nature of Planetary Orbits.

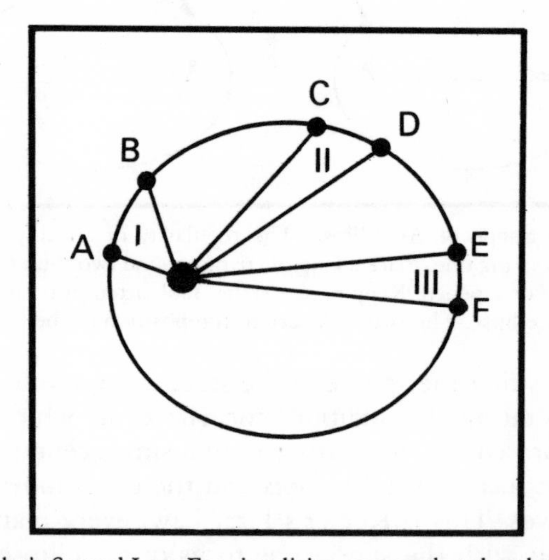

Figure 1.5. Kepler's Second Law. For simplicity, assume that the planet (represented by the smaller black dot) takes one month to move from position A to position B. The area enclosed between A, B and the sun (marked I) has the same area as that contained in the shapes II and III. The time required for the planet to move from C to D, or from E to F, will therefore also be one month. The speed of planetary movement slows as the distance from the sun increases.

more slowly; when it is close to the sun, it moves more rapidly. This, Kepler's Second Law, is illustrated in Figure 1.5.

Kepler also found that there was a simple relationship between the average planet-sun distance (semimajor axis)[6] and the time the planet takes to go around the sun. This Third Law is a quantitative relationship: the cube of the semimajor axis divided by the square of the length of time required to complete the orbit is the same for every planet.

Here Kepler had his truly perfect and mystical relations. These were undoubtedly wrought by God since no mere chance could have produced them. Kepler had "understood" everything, so he then wrote a book, *Harmonice Mundi, (The Harmony of the World).* In the introduction, Kepler, at the same time incredibly modest and remarkably presumptuous, says:

> I mockingly defy all mortals with this open confession: I have robbed the golden vessels of the Egyptians to make out of them a tabernacle for my God. . . . If you forgive me, I shall rejoice. If you are angry, I shall bear it. Behold, I have cast the dice, and I am writing a book either for my contemporaries or for posterity. It is all the same to me. *It may wait a hundred years for a reader, since God has also waited six thousand years for a witness. . . .*[7]

Kepler was correct in guessing he might have to wait for an understanding reader. No one in his age, including his more famous contemporary Galileo, did understand what he had achieved.

Galileo: Embattled Scientist, Propagandist, Martyr

Galileo was everything that Kepler was not, and Kepler was many things that Galileo could not be. Kepler was a passionate mystic and an obscure mathematician whose quantitative insights made it possible for Newton to explain gravitation. Galileo was a simple realist, logical, with a passion not so much for understanding (he already knew) as for convincing others. Kepler was the model of the recluse scientist even though his disposition was incompatible with pure reasoning. Galileo with his straightforward logic was the prototype of the scientific reformer.

[6] The semimajor axis is half of the longest diameter of the ellipse. (See Figure 1.4.)

[7] Emphasis added.

At a relatively early period, Galileo was convinced that the Copernican theory was correct, but he needed more tangible evidence. For him, seeing was believing, so Galileo reinvented the telescope for himself and became the first man to aim the telescope at the heavens. He looked into the night sky and saw the mountains on the moon. He looked at Jupiter and saw moons circling around it, a small version of the solar system. He observed the phases of Venus and argued that this was proof that Venus orbited the sun and not the earth. He also noticed through his telescope (and this is not directly connected to the arguments of the Copernican system) that the Milky Way appeared as many stars which the naked eye could not distinguish. A milky background remained even through the telescope, which hinted of ever more celestial objects. Thus Galileo had the first indication of the nature of our galaxy: an assembly of an enormously huge number of heavenly bodies. Galileo wrote these unusual observations down in a book called *Sidereus Nuncius*.[8]

Yet Galileo made some basic scientific errors in his attempts to establish the sun as the center of our system. Galileo was backing the orthodox Copernican theory, based on circular orbits and the consequently complicated epicycles. He held this view in spite of Kepler's efforts to convince him of the great simplification that elliptical orbits provide.

Galileo was a master of debate and was very successful in demolishing the arguments for the earth-centered universe, making many enemies in the process. However, he was never able to furnish clear physical proof that the earth orbited the sun. Because of this, he couldn't displace Tycho de Brahe's alternative to the Copernican view. In fact, de Brahe's model had two advantages. It explained the absence of stellar parallax (while the Copernican view did not), and it required no Biblical reinterpretation. So de Brahe's alternative was supported by most of the scientifically competent members of the Church.

On one essential point, Galileo, an excellent physicist, went to the heart of the problem raised by Copernican theory. If the earth is moving and rotating, why don't people notice any motion? Remarkably enough, this very obvious question had not been addressed in any earlier writings. Galileo not only asked the question but helped to answer it, establishing a principle to which I shall return in even more detail when Einstein's work is considered.

[8] The original title is sometimes rather horribly translated (though accurately) as *The Starry Messenger.* I prefer to call it *A Message From the Stars.*

The principle is that under some circumstances, it is not possible to know whether one is moving or at rest. Enclosed in a box moving at constant velocity along a straight line, one would feel the same as if at rest. A smooth airplane flight is a practical example for some people. One's observations would not be sufficient to determine whether the box or the surroundings had moved. Galileo talked about a uniformly moving ship and of dropping an object from the mast of such a ship. The object dropped will land at the base of the mast regardless of whether the ship is moving uniformly or at rest. He thus established with fair clarity what is today known as the principle of inertia. He understood straight-line motion and the accelerated motion of a body under the influence of the earth's gravitation.

Nevertheless, Galileo's ideas about the motion of bodies were not complete. He still believed that the natural motion of an object around the earth was a circle. Considering this, the mistakes he made do not seem so extraordinary. He did not apply the simple principles of physics that he found in his workshop to the universe as a whole. Galileo was primarily interested in proving that the earth moved around the sun. His contribution to the heliocentric theory was to start the process which reconciled the earth's real motion with people's failure to notice its very great speed.

Kepler suggested that tides are due to the influence of the moon. Galileo dismissed this as astrological superstition and came up with his own (wholly incorrect) theory.[9] Galileo used his theory of tides as proof of the Copernican view. His statement that the orbital motion of the earth would produce the tides, apart from its other faults, even contradicted his own correct principle that the essentially uniform motion of the earth cannot be detected within the moving system. So his "proof" was clearly wrong.

However, Galileo was so determined to convince everyone that the Copernican theory (and his own views) were correct that in 1630 he decided to publish a work then titled *Dialogue of the Flux and Reflux of the Tides*. There were four parts to the *Dialogue* corresponding to four days of discussion. The fourth day deals with Galileo's incorrect theory of tides. On the whole, though, the thrust of the treatise was right. It explained why the motion of the earth could not be easily

[9] Galileo's theory of tides was based on the idea that the earth's rotational velocity (about its axis) and orbital velocity (about the sun) would add during the night and subtract during the day. This would cause the land to move more rapidly at night than during the day. The waters of the ocean could not follow this motion, and so one high tide would occur around noon each twenty-four hours. The facts, that there are two high tides each day and that the hour of their occurrence changes from day to day, Galileo credited as secondary effects without any detailed explanation.

detected. However, the scientific principles in this *Dialogue* were much overshadowed by the controversy which its eventual publication engendered.

The contemporary Pope, Urban VIII, was a remarkable man: intelligent and considerably knowledgeable about science and astronomy. He had been Galileo's friend and defender for a number of years.[10] A few years earlier, Galileo had conferred with him about the defense of the Copernican theory. Urban had cautioned Galileo to expound his views *ex hypothesi*, merely as interesting hypotheses. In that way he could avoid running afoul of official Church policy. Urban did Galileo another, perhaps bigger, favor by not accepting his ideas about tides. He suggested that Galileo's work should have a title not related to the tides. Galileo found a much better title: *Dialogo . . .sopra i due Massime Sistemi del Mondo,* which can be translated as *Dialogue on the Two Great World Systems* (meaning the Aristotelian and Copernican systems).

Galileo had other friends, both colleagues and churchmen, who had given him similar cautionary advice: argue for the Copernican view as a simplifying hypothesis. The Church at this time was amenable to reinterpreting relevant Biblical passages, provided conclusive proof was presented. This had been done on other occasions, most notably in regard to the shape of the earth. The Church took the position in this instance that the Bible had figurative rather than literal meaning.

Before publishing his *Dialogue*, Galileo took the manuscript to Urban VIII. Unfortunately for everyone, Urban was too busy to read it, and the task fell to the Chief Censor and Licenser, Father Niccolo Riccardi. Riccardi was a well-meaning man and a friend of Galileo's supporters. But, in addition, he could not understand the contents of the *Dialogue.* He knew that Urban VIII was a great admirer of Galileo and had basically agreed to the idea of the manuscript; he was under considerable pressure from Galileo and his friends; and so he eventually allowed his office to pass the manuscript for publication.

Now what was most remarkable, and what undid Galileo, was not so much the contents but the tone of the *Dialogue.* In this treatise, two intelligent people, Salviati (representing Galileo) and Sagredo (a neutral bystander), discuss Galileo's views with a third, Simplicio. The name of the last is descriptive enough. Simplicio is always the slowest to understand any argument. Yet Galileo put the arguments of Pope Urban, his friend, into the mouth of Simplicio. Furthermore,

[10] When he was Cardinal Maffeo Barberini, Urban had even written an ode to Galileo, *Adulatio Perniciosa.*

at the end of the *Dialogue*, where the incorrect theory of the tides is being explained, Simplicio objects strenuously; Salviati does not accept his arguments but only says, "If God wills that Simplicio be right, then he will be right because God is all powerful." This type of allusion could not possibly have pleased Urban VIII, and it didn't.

Within a few weeks of the publication of the *Dialogue*, Urban VIII realized that Galileo had made him and his Holy Office look foolish. Galileo's book was confiscated, and he was summoned to appear before the Inquisition in Rome. Galileo was not in a comfortable position, and he did not want to go. He pleaded old age and illness, but eventually he had to face his accusers. An additional problem was that in their earlier discussions, Galileo had asked about the decree of 1616, which prohibited arguments for the Copernican view without fair presentation of the counter-arguments. Urban had flatly refused to revoke it. Could Simplicio's remarks throughout the *Dialogue* stand as fair counter-arguments?

To appreciate the situation fully, one must look at both sides. Pope Urban VIII had every reason to object to the way Galileo had sidestepped his advice and the authority of the Church. The Church was the establishment. As unlovely as this word sounds today, it does carry the implication of responsibility. Tycho de Brahe's theory continued to have great authority with the Jesuit astronomers. Urban might have ordered a reinterpretation of the Bible, but because there was the absence of a conclusive proof of the Copernican theory, the Church might later have had to revert back to an earth-centered universe. The Pope had to consider the effect of that possibility on the authority of the Church and the confidence of the people in it. Because of the provocative nature of his attack, it was not sufficient for Galileo simply to say: "I don't know for sure."

Of course, the essential truth was eventually found on Galileo's side: the earth does sail around the sun. In the confrontation between Galileo and Urban VIII, it is fair to say that neither was completely right or wrong. Both were clearly men of great ability and, also, of great vanity. The myth is that the hidebound, anti-science, prejudiced Church forced one of the greatest of scientists to denounce his own teachings. Reality is often more interesting than myth.

The immediate results of this confrontation also deserve to be examined. Apart from Galileo's being spared the embarrassment of having his incorrect tidal theory widely circulated (because the book was banned), he was also forced by a none-too-painful house arrest to do something that great scientists have occasionally neglected to do. Not being able to campaign for his astronomical views, he wrote down much of the excellent work he had done in physics in previous years. His great fame is justified and enhanced by this work rather than by his painless martyrdom.

Smoother Pebbles, Prettier Shells: Newton's Apple of Simplicity

Galileo's story has considerable drama, but it lacks the real simplicity I would like to demonstrate. While Kepler's formulation of planetary motion summarizes a vast array of observations in a simple form, the information still does not look truly simple or logical. Why are there ellipses? Why do the mystical formulas exist?

The suspected hundred-year-wait for a reader was finally over when Isaac Newton picked up Kepler's book and used it to put the capstone on the Copernican revolution. Newton was the first to understand and prove in a consistent manner that the same laws hold on the earth and in the heavens. The idea of a universal natural law had been slowly emerging, but Newton turned it into an indisputable scientific fact. The most important part of Newton's work was done when he was a young man studying at Cambridge. A plague had closed the university, and Newton went home to Woolsthorpe. It was here that the apple figuratively fell on his head. The apple symbolizes an important step toward a new point of view which today tends to be taken for granted.

Newton's insight may best be understood by starting from Kepler's Third Law which describes the simple relationship between the semimajor axis of the ellipse and the time the planet needs to go around the sun. This can be written as a formula: a^3/t^2 = constant, where a is the length of the semimajor axis of the orbit, and t is the length of time required to complete one revolution. Newton's contemporaries knew that this law held when the orbits of planets were considered and also when comparing the orbits of Jupiter's moons, though in this case, a different constant has to be used.

Newton realized that the same law should apply for the motion of the moon around the earth and the motion of an object hurled with the appropriate velocity around the earth. Both had elliptical orbits. At that time, an experiment was not even remotely feasible, but today people have done this very thing using satellites instead of apples. The apple was Newton's thought-satellite. Newton could touch the apple, and today man has touched the moon.

Before Newton, the apple and the moon belonged to two different worlds: the apple was terrestrial, the moon celestial. Since Aristotle's time, the natural state of motion for objects in the heavens was perpetual circular motion, and the natural state for objects on earth was one of rest. Kepler had demonstrated the ellipse, and Galileo had correctly argued that the state of uniform motion was the correct one for objects on earth, but even for them, the celestial and terrestrial laws were different. Perhaps the greatest of Newton's contributions was to point out the universality of physical laws—the same laws hold for both the heavens and earth. Kepler's celestial laws, when applied to an apple falling on earth, predict the motion which is actually observed. This was certainly a giant leap forward in the pursuit of simplicity.

In the course of unifying heavenly and terrestrial laws, Newton gave a simple, enduring formulation to mechanics. A change in velocity had to have a reason—hence the idea of *force.* Newton applied this idea—in the spirit of his times—in similar ways to quite different causes, such as the action of a spring, gravitation as seen on earth, and gravitation between heavenly bodies.

The logical (and boring) formulation of force is $\mathbf{F} = \mathbf{ma}$, force equals mass times acceleration. If one wishes, one may say mass is defined in this manner, provided one already knows what force is. One might prefer to assume that mass is the known quantity and define force by this equation. What is seen here (and will be seen again in the following chapters) is the establishing of relations between various concepts in order to define terms. Which comes first and which comes later is often a matter of choice. What is obvious is that an object with twice the mass of another requires twice the force to change its state of motion.

Kepler's Second Law (equal areas are swept in equal times for a single planetary orbit) could be explained by assuming that the sun attracted planets with a "central" force, that is, a force pointing from the planet straight toward the sun. (See Appendix II.) Kepler's Third Law (a^3/t^2 = constant) could be explained by assuming that at two times the distance, the force was diminished fourfold, and that in general an inverse square law held: the force upon a given planet is a constant divided by the square of its distance from the sun. (See Appendix III.) The constant in turn is proportional to the mass of the attracting object (in the present case, the sun).

Having made this step, Newton could conclude that the gravitational force of the earth, or the sun, produced the curving motion of projectiles on earth and in the orbits of moons and the planets. Newton apparently realized all this during his enforced holiday at Woolsthorpe.

Newton seems to have been very reluctant to publish. Some of the ideas which are considered parts of Newton's theory were generally discussed and accepted during this period and may have occurred to several of his contemporaries independently. Robert Hooke, an important scientist at this time, is remembered today for Hooke's Law: the deformation of an elastic body is proportional to the applied force. This may be a rather inadequate monument for a man of real scientific stature in his day. The confusion about credit exists because of Newton's reluctance to claim his discoveries and, in general, get into scientific controversies. His abilities were clearly recognized though. He returned to Cambridge from Woolsthorpe and became a professor, but for about twenty years not much was heard of his tremendous discoveries. Then a remarkable incident occurred between Newton and Edmund Halley.

Both Kepler's laws and the inverse square law of gravitation were widely known in this period. Halley was interested in comets. At that time, comets were thought to come in from and return to infinity (have parabolic orbits). Halley had studied the history of comets and thought he had detected some periodicity.[11] Halley suggested to Newton that the orbits of comets might be very elongated Keplerian ellipses.

Halley had a pointed question to ask Newton: It is generally believed that Kepler's First Law (that planets move on elliptical orbits with the sun at one focus) follows from the inverse square law; can this be proved? Newton replied that he had obtained that result many years earlier while he was at Woolsthorpe. Halley asked for the derivation. Newton, not having considered the problem for many years, was unable to reproduce it.

The story is that it took Newton two (what must have been very frustrating) weeks to reproduce the result. Then, wanting to be very sure that the proof wouldn't be forgotten again, he published *Philosophiae Naturalis Principia Mathematica*, his justly famous *Mathematical Principles of Natural Science*. (See Appendix IV.) This book contains a lot of precise and very important ideas, including not only Kepler's laws and the correct theory of tides but many other beautiful applications. With its publication, all doubt as to the correctness of his approach disappeared.

Another of Newton's remarkable accomplishments should be specifically mentioned: the gravitational forces emanating from a

[11] His most famous prediction from this study was the return of a comet, well after his death. That comet, last seen in 1910 and expected back in 1986, now bears his name.

spherical shell of uniform density behave as if the whole mass of the shell were concentrated at the center, provided the gravitational force is measured from outside the shell. Today this remarkable result can be explained in very simple terms. Newton may have found the result by more complex means.

The idea that is available to us today was introduced more than a century after Newton by another great British scientist, Michael Faraday. This is the concept of *lines of force.* These lines of force pictorially indicate the direction in which the force acts, and the density of the lines (the number of lines crossing a unit area) gives us the strength of the force. Imagine a multitude of "spokes" emanating equally in all directions from a mass concentrated in one point. At this center, these lines are so dense that they (roughly speaking) may be said almost to touch. Imagine spheres around the center at varying distances. The density of spokes will decrease in each successive sphere, even though the number of spokes stays the same. (See Figure 1.6.)

The surface area of these imaginary spheres can be found using the formula $4\pi r^2$. As r, the radius (which is also the length of the imaginary spoke) increases, the area grows larger. The number of lines of force remains the same, but the density of the spokes decreases as the inverse square, and so does the force of gravity. This approach provides a very clear pictorial illustration of the inverse square law.

Assume that the number of lines of force emanating from the mass point (the point at which the mass is centered) is proportional to the amount of the mass. Assume also that these lines do not end anywhere but go outward to infinity. Finally, assume that the gravitational force points in a reverse direction to these lines. Imagine a hollow sphere with mass distributed uniformly on the surface, and take the perspective of a point located outside the sphere.

One can see that the forces acting on this point in the region outside the shell cannot depend on whether an equal mass is distributed over the shell (if this distribution is uniform) or whether the mass is all located at the center. The lines of force appear the same in either case since their number is proportional to the amount of mass, regardless of its distribution.

Inside the shell, the spherical symmetry of the mass prevents the lines of force from pointing in any direction except radially. Because these lines can originate only in a mass, and because there is no mass in the center of a hollow shell, and because these lines are supposed

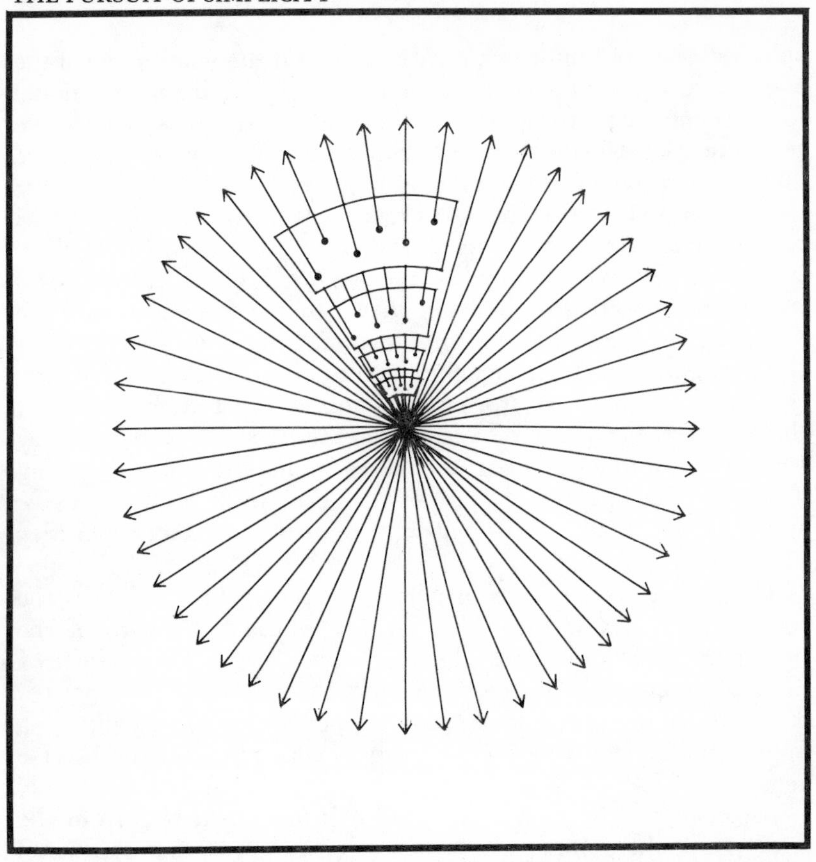

Figure 1.6. Pictorial Representation of the Inverse Square Law of Gravitation. The gravitational force decreases as the density of the lines decreases, (which is hard to see in two dimensions), or to state the law in its usual form, gravitational force decreases as the square of the distance from the gravitating body. The portion of the surface area of imaginary spheres within spheres enclosing six lines of force can be seen to increase as one moves outwards from the central mass, thus decreasing the density of the lines.

to go to infinity, there can be no lines of force inside the shell and therefore, no gravitational forces.

It is doubtful that this pictorial demonstration was present in Newton's thinking, but another cornerstone of modern mathematics and physics was his point of departure. Newton formulated the idea that the motion of particles must be followed during very, very short time intervals. This insight was surely one of the greatest single advancements ever made by any man. It is the basis of differential equations. What Newton proposed (and more or less simultaneously and independently, Gottfried Leibnitz and Isaac Barrow suggested) was that instead of trying to work on large segments of motion at once, such as a complete orbital movement of a planet, one should look at the tiniest imaginable segment and then at the next tiny segment, continuing until their totality describes the entire movement. Thus, what would hardly have been possible to describe mathematically on a holistic basis became demonstrable when taken in the smallest pieces.[12]

Newton's great contribution was combining mathematical ideas with physical formulations to complete the picture. He brought the light of understanding to Kepler's laws. He made immense contributions to the study of optics. Yet, Newton saw the world very differently than his predecessors. Johannes Kepler, in his immodest modesty, compared his discoveries of God's laws to God's work. The passage from Newton's pen that I like best is:

> I do not know what I may appear to the world, but to myself I seem to have been only like a boy playing on the seashore and diverting myself in now and then finding a smoother pebble or a prettier shell than ordinary, whilst the great ocean of truth lay all undiscovered before me.

To me, Newton's pebbles and shells are indeed smooth and pretty because they are so very simple. However, they did not seem simple before he recognized their nature. Mankind shall find more pebbles and more beauty.

The ocean of truth now offers a vista of a hundred billion suns clustered in the galaxy Galileo glimpsed with his telescope. The universe contains billions of galaxies which, having started 15 billion years ago in a small space, now are moving apart at tremendous speeds. Newton pointed out that the laws that apply to galaxies are the same as the laws that apply to atoms. Regardless of how different galaxies, atoms and the application of laws may superficially appear, their common consistency is now clearly visible.

But the simplicity in the structure and origin of the universe has not as yet been fully discovered, and therefore all people can repeat with Newton, the great ocean of truth is still undiscovered. That ocean is incomparably greater than Kepler, the mystic, dreamed. It is more complex than the logical mind Galileo imagined. But Newton, in his true humility, may have had a glimpse of it.

[12] What are these smallest pieces? One can always think of pieces which are even smaller. However, one can raise the question: what are the properties which can be satisfactorily defined in the limit when these pieces approach zero? In general, velocity is distance divided by time in a uniform motion. In differential equations, velocity is still distance divided by time provided both of these approach zero.

Chapter Two

THE GEOMETRY OF SPACE AND TIME

The topic of this chapter, Einstein's theory of relativity, is typical of the ideas developed in science near the beginning of this century. These theories contradict accepted ideas—they even seem to contradict common sense. As a result, most people have given up the effort to understand them.

As the reader will see, relativity is not difficult because of its mathematical complexity—a knowledge of high school mathematics suffices. The difficulty in understanding lies in the novelty of the concepts. Overcoming amazement requires imagination and care. One quick reading will not suffice, but perseverance will provide a surprising, correct and simpler view of the world.

Getting Relatively Coordinated

I have claimed that Pythagoras was the first scientist. (Incidentally, he was also the first scientist to get into trouble by meddling in politics.) A modern consequence of his theorem became important because of an invention of the French mathematician and philosopher, Descartes. In a successful attempt to reduce geometry to straightforward mathematics, Descartes invented what is called a *coordinate system.*

The simplest form of a coordinate system is two perpendicular lines called the **X** and **Y** axes. The position of a point (p) is then described in the following manner. (See Figure 2.1.) Starting from the intersection of the two axes, called the origin (0), proceed along the **X** axis for a distance (**x**). Thereafter, move upwards (parallel to the **Y** axis) a specific distance (y). The point is described by the lengths one moved along each axis. The numbers, **x** and y, are called the coordinates of the point, and the axes are called a coordinate system. (See Appendix V.)

Finding a point using the coordinate system is like finding an address in a well-laid-out city. For example, in New York, it is not difficult to understand how to go to the intersection of 2nd Avenue and 44th Street. The idea of a coordinate system is so simple that I

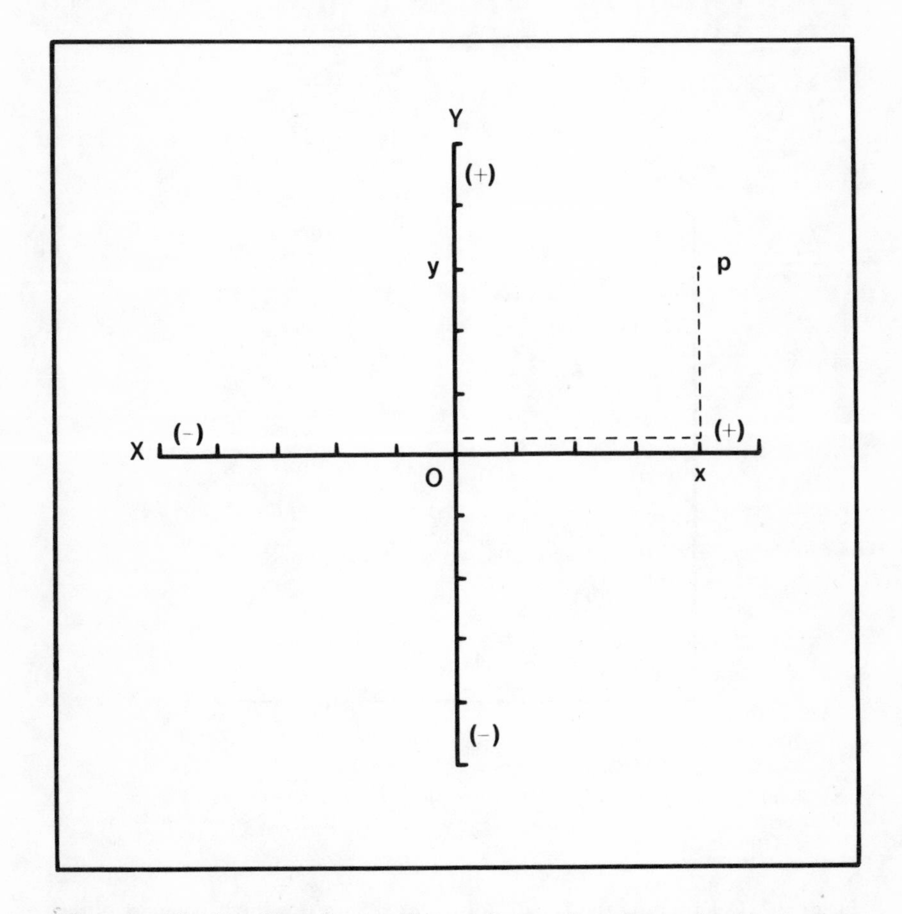

Figure 2.1. The Coordinate System. The coordinates of a point may be either positive or negative. In the figure above, the coordinates of p are $x = +4$, $y = +3$.

am tempted to apologize for explaining it. Yet it has become an essential part of the vocabulary of mathematics and physics.

A more modern idea—that of an invariant—may appear even simpler, but its applications have been numerous and surprising. An invariant is a quantity which under certain conditions does not change. The relevant beginning of the discussion of an invariant results from a joint application of the ideas of Pythagoras and Descartes.

Does knowing the coordinates **x** and **y** of a point make it possible to determine the distance of this point from the origin? By inspecting Figure 2.2.,one can see that a right angled triangle can

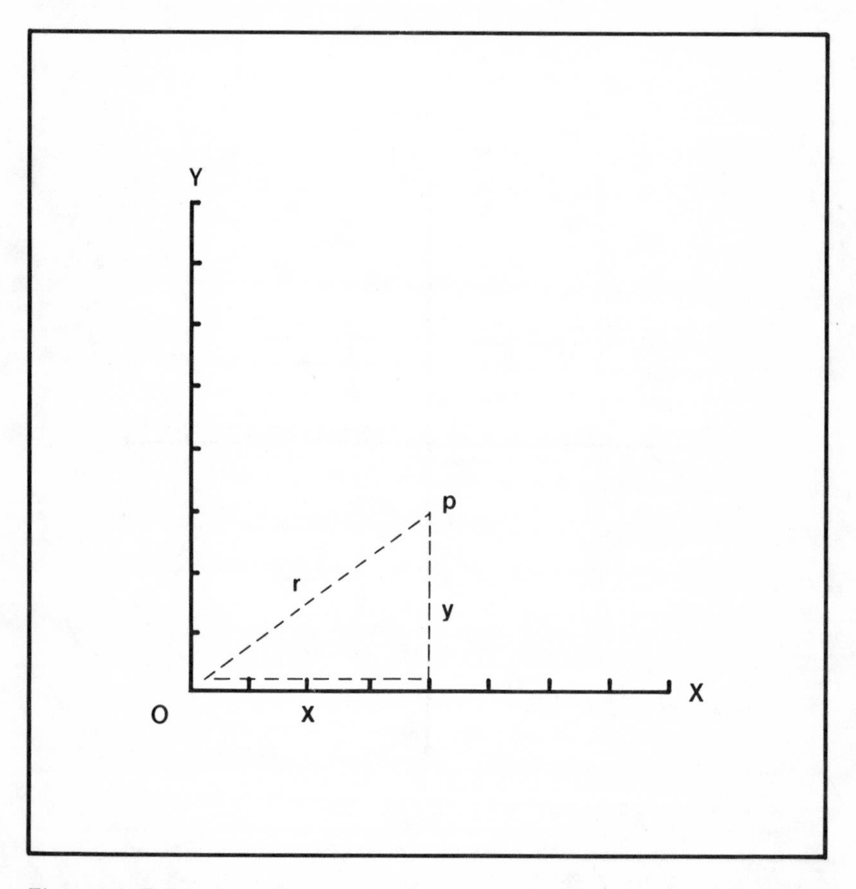

Figure 2.2. Determining the Length (distance) r. Since r is the hypotenuse of the right angled triangle, $x^2 + y^2 = r^2$, or in this case $4^2 + 3^2 = r^2$. The distance of p from the origin is therefore equal to 5, the square root of the sum of the squares.

easily be formed. Follow the path used to reach **p** (move along the **X**-axis; then move parallel to the **Y**-axis). After reaching **p**, draw a straight line from **p** to the origin. The straight line of length **r** is the *hypotenuse*. According to the Pythagorean theorem: in a right triangle, the sum of the squares of the two shorter sides is equal to the square of the longest side, or in mathematical shorthand, $r^2 = x^2 + y^2$.

In two-dimensional space, the idea of an invariant arises when one considers another coordinate system with the same origin. (See Figure 2.3.) The new coordinates will be called **x′** and **y′** and the new axes **X′** and **Y′**. Since the distance of **p** from the origin remains the

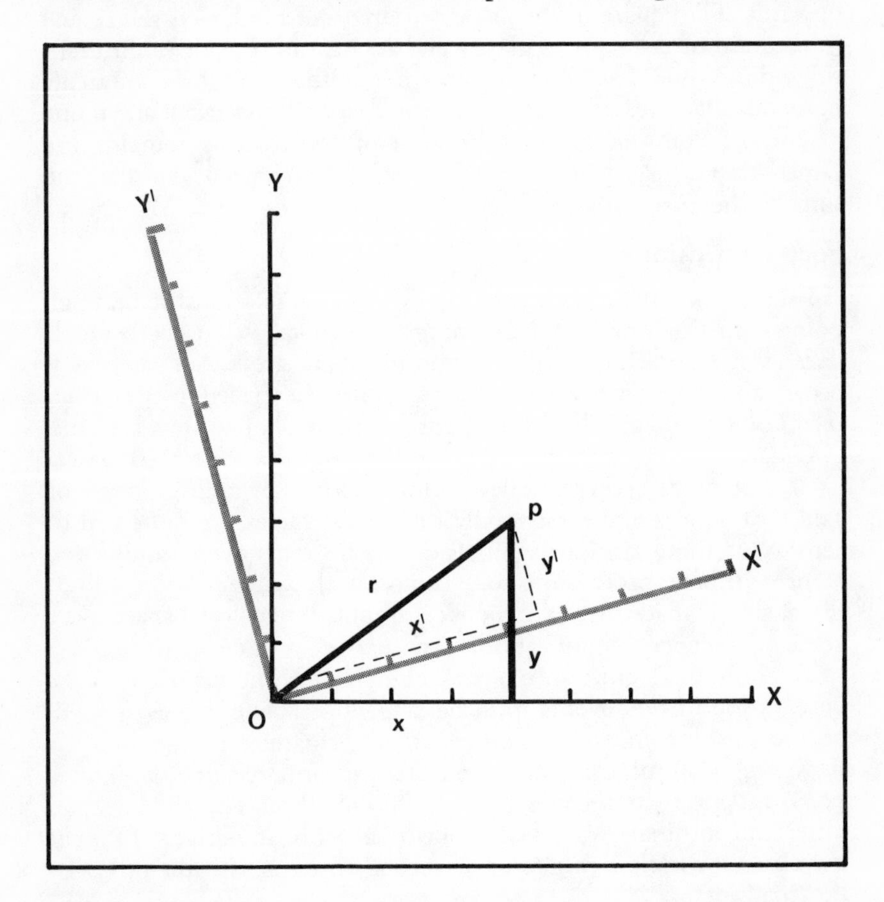

Figure 2.3. The Distance *r* in Rotated Coordinate Systems with a Common Origin. In two- or three-dimensional coordinate systems which have a common origin, the length of *r* does not vary.

same in the new system, **r** continues to be called **r** rather than **r′**. In both systems, regardless of their orientation, the distance of **r** from the origin (represented by the diagonal line) is the same.

Since **r** squared is equal to the squares of the values of either set of coordinates, $r^2 = x^2 + y^2$ and $r^2 = x'^2 + y'^2$, the sums of the squares from the two coordinate sets must also be equal. One may now say that in two-dimensional coordinate systems with a common origin, the sum of the squares of the coordinates is an invariant; it does not change its value when it is compared in two rotated systems.

It is not hard to go from here to three-dimensional space. Three coordinates, **x**, **y** and **z**, may be drawn toward the north, east and upwards. The square of the distance (length of line), r^2, is then equal to the sum of the squares of **x**, **y** and **z**. One can consider a different three-dimensional coordinate system (so that the three mutually perpendicular axes do not point north, east and upwards), and if one assumes a common origin, the sum of the squares remains the same—that is, $x^2 + y^2 + z^2 = x'^2 + y'^2 + z'^2$. The invariant is still r^2, the sum of the three squares.

Four Dimensions?

I shall now shift from common geometry—a relationship between points—to the world of happenings or events. What the event is doesn't matter. Rather, the important ideas are *when* the event occurs and *where* it happens. The position is described in the earlier terms of **x**, **y** and **z**. When it happens will be symbolized by **t** for time.

Consider two events: one occurs at the origin ($x = y = z = 0$) and at $t = 0$. The other occurs at a new point described by a different set of values of **x**, **y**, **z** and **t**. Generally these new values are not equal to zero. Now there are four symbols describing each event, and if one wanted to have pretensions, one could say this is dealing with four-dimensional space. Actually the idea of four-dimensional space has a somewhat deeper significance.

To show the significance of four dimensions, the invariant in the observation of two events must be determined. The two events will be the pushing in and popping out of a cigarette lighter in a car traveling at 60 miles per hour in a straight line. Assume the lighter pops out after 15 seconds. Two different observers will use two different coordinate systems. Let one observer be the driver of the car traveling at a mile a minute. The second observer is a hitchhiker by the roadside.

First, consider the distance between the two events in space. From the driver's perspective, both events occur a foot and a half in front of and slightly to the right of him. The two events are not separated in space. For the hitchhiker, however, the two events are clearly separated by a quarter of a mile. Are these two points of view of a distance to be considered as equally justified? Isn't it more sensible to take the point of view of the hitchhiker who is at rest? Isn't it obvious that the two events take place a quarter of a mile apart? This would be obvious only if it were truly clear who is at rest and who is moving.

In the previous chapter, I discussed a law of physics originally formulated by Galileo. Using the example of the object falling to the base of a mast regardless of whether the ship is moving uniformly or at rest, Galileo stated that it is not possible to distinguish between a state of uniform motion and a state of rest.[1] This means that if one moves with uniform speed, one is restricted to making observations only from within one's own coordinate system. In this case, the automobile is that coordinate system, and its slightly uneven motion is to be neglected. The fact that it took humans so long to recognize the earth's great motion is strong evidence for the correctness of Galileo's principle.

The problem of two observed distances between two events might be solved by adopting a democratic perspective. Most people are at rest while cars move in all possible directions. Leave the decision to the majority who are "at rest." Before the Copernican revolution, this point of view was adopted as the obvious one: the earth was at rest. But today the situation is different. One knows that an observer who is in Australia[2] will be moving in a direction quite different from one's own because of the rotation of the earth. Further, an observer at rest with respect to the sun would see the whole earth moving by at some 17 miles per second. Another observer at rest at the center of our galaxy would see the whole solar system move along at a velocity higher by another factor 10. Still another observer sitting in a distant galaxy would see the entire Milky Way moving at perhaps 20,000 miles per second.[3] Who is moving? Who is at rest?

An important consequence of the Copernican revolution was that when the earth became a relatively tiny rotating ball sailing through

[1] Although Galileo did not apply the principle in his tidal theory, he correctly stated that uniform motion cannot have a noticeable effect if observations are confined to a uniformly moving system. Galileo did not recognize the conflict because for him terrestrial and heavenly laws were not merged into the concept of laws of nature.

[2] With apologies to any of the 12 million Australians who happen to read this book.

[3] The actual speed could be any of the physically possible speeds, depending upon the galaxy from which the observer chose to view the event.

space, the idea of a body at rest was lost. This fact was first suggested by Galileo and was later incorporated into a mathematical-physical system by Newton. Galileo's law can be restated: all uniformly moving coordinate systems are equivalent; they are equally valid. This, as is now obvious, leads to the conclusion that the distance between two events (or more specifically, $x^2 + y^2 + z^2$ which is denoted as r^2) is not an invariant! *The distance between two events depends on the observer.* In this regard, the geometry of space and time is quite different from the geometry of space alone. Adding the fourth-dimension leads to perplexing consequences.

The ideas necessary for this conclusion were available at the time of Galileo. But even after Galileo, there remained a "classical" invariant: the time elapsed between two events. In this example, that fifteen seconds elapse between pushing in the lighter and its popping back out seems to be something on which the driver and hitchhiker could agree. As a matter of fact, Newton clearly stated that time is an invariant. It remained an invariant until 1905. Einstein found a flaw in this idea.

Finding the Invariant: Where is the Anchor?

Perhaps it would be helpful to my readers to explain that at this point I shall deviate from the scientific principle of orderly deduction in the presentation of a theory. I am going to use what I hope is a more easily understandable procedure which I call the Grasshopper Principle. I jump to a conclusion and then ask, "Where am I?" I hope it will be easier for the reader to find a whole new set of unusual facts than to be confronted with them piecemeal. In the Grasshopper method, the reader does not have to deal with the question of which old "obvious" facts still remain valid.

The invariant has the role of a fixed anchor in a sea of relativity. The word *relativity* is a bit unfortunate because in Einstein's theory, the main point is not what is relative, as most people believe, but rather what is invariant, unchanging, and hence not relative at all. Suspecting that neither the distance nor the time between two events was invariant, Einstein went on to provide a quantity that is an invariant. To understand it, remember first that each observer has a coordinate system. In the observation of two events, three factors must be considered: the speed of light,[4] denoted as c, the time

[4] The reason one must consider light is that light transmits messages from one coordinate system to another. The fundamental role of the speed of light is by no means obvious. Its introduction is the main point in Einstein's accomplishment. I recommend that the reader accept this statement as a jump of the grasshopper. S/he will have plenty of occasion to find out the consequences while reading along in the chapter.

elapsed between the two events (t), and the distance between the two events (r). In particular, one should notice that the quantity ct—the product of light velocity and time elapsed—actually represents the distance light will travel in the time between events.

If one squares the distance light travels in the time between events and subtracts the square of the distance between the two events, the result is the new invariant adopted by Einstein. It does not vary—regardless of the coordinate systems from which the two events are observed. The quantity $(ct)^2 - r^2$ is the new anchor of certainty that all can share, driver, hitchhiker, and everyone else.[5]

Changing the concept of time in any sense is, of course, an insult to one's basic intuitions. But because the value of the velocity of light (186,000 miles per second) is very large, the square of ct is much larger than the square of r for most pairs of events. This, as presently will be seen, prevents one from noticing in most practical cases the proposed peculiar behavior of time.

Consider the case of the car lighter again. From the perspective (coordinate system) of the driver,

$$(1)\ (ct)^2 - r^2 = (186{,}000 \cdot 15)^2 - 0^2.$$

For the hitchhiker, a different coordinate system exists, so the distance in this case is r′. The hitchhiker's r′ is one quarter mile (1/4). The invariant for the hitchhiker would be calculated as

$$(2)\ (ct\ )^2 - (1/4)^2 = (186{,}000 \cdot t')^2 - 1/16.$$

According to Einstein, $(ct\ )^2 - r^2$ is an invariant, that is,

$$(ct\ )^2 - r^2 = (ct')^2 - r'^2.$$

Therefore, (1) must be equal to (2) or

$$(186{,}000 \cdot 15)^2 - 0^2 = (186{,}000 \cdot t')^2 - 1/16.$$

This works out that t′ is equal to 15.00000000000006 seconds.

It is not surprising that the hitchhiker and the driver did not notice the difference in their perception of time. For practical purposes, $t = t'$. Apparatus sensitive enough to detect this extremely small difference between t and t′ may well not be built in the next thousand years. That t and t′ are not the same will be felt as absurd. (But in this case, only slightly absurd?) In principle, however, the measurement could be made, and one could find out directly whether or not Einstein was right.

[5] The non-Grasshopper approach is less simple but more straightforward. I offer a note at the end of this chapter to explain *how* time and distance change when viewed from two systems moving relative to one another. The readers who wish to use more mathematics in the examples offered in this chapter will find the necessary information to prove these incredible ideas by yet another means by referring to this section.

The reason that t and t' differ so little is that the automobile can move only very slowly compared to the velocity of light. So instead, consider an astronaut[6] who is moving with half the velocity of light ($c/2$). The astronaut pushes in the lighter while passing Mission Control. In this case, because the distance between events (r) is again zero, the invariant from the point of view of the astronaut is,

$$(ct)^2 - r^2 = (186{,}000 \cdot 15)^2.$$

For Mission Control, the distance between events (r') is 15 seconds times $c/2$, or $r'^2 = (186{,}000/2 \cdot 15)^2$. The calculation to find t' using Einstein's invariant is then

$$(186{,}000 \cdot 15)^2 = (186{,}000 \cdot t')^2 - (93{,}000 \cdot 15)^2.$$

This calculation when completed shows that t' is approximately 16.77 seconds. The difference from 15 seconds is easily detectable.

Historically, the story I have presented is false. Einstein did not introduce the invariant $(ct)^2 - r^2$. This was done a few years afterwards by Hermann Minkowski, a mathematician, who thereby established a beautifully simple form of Einstein's original ideas. Einstein actually started by proposing a more basic invariant. He stated that the speed of light is the same for all observers. At first, this does not seem startling, but when one tries to use this idea, it contradicts common sense.

Think again of the astronaut and Mission Control. In this case, assume that there is a satellite exactly 186,000 miles from the earth. (This makes it a little closer than the moon.) The astronaut is flying towards the satellite at half the speed of light. Just as the astronaut passes Mission Control on earth, Mission Control sends a laser beam light flash to the satellite.

The two events to consider are the firing of the laser on earth and the arrival of the light flash on the satellite. Viewing these two events from the perspective of Mission Control, one can write down the expressions for Einstein's (or more properly, Minkowski's) invariant. In the case of Mission Control, the laser flash took one second to arrive at the satellite, so the invariant formula is

$$(186{,}000 \cdot 1)^2 - (186{,}000)^2.$$

This means that the invariant equals 0.

For the astronaut, the first of the two events—the firing of the laser—occurs (as in the case of Mission Control) at the origin of the coordinate system because at that time he passes Mission Control. Therefore, for both observers, space and time can be counted as 0, at the origin. Considering the second event, the arrival of the laser

[6] Astronauts move much faster than race drivers. Comparing the velocity of either with the velocity of light still leaves the speed of man's motion negligible. However, with an astronaut, I am allowed much greater license.

beam on the moon, the time will be called t' and the distance r'. It is clear that $(ct')^2 - r'^2$ must equal zero, the same as was calculated by Mission Control. From this it follows that $(ct')^2 = r'^2$, or $c = r'/t'$. In this case, the measured distance (r') divided by the measured time (t') has the same value (c) for the man at Mission Control and for the astronaut. Therefore, the speed of light is constant. This result, at first thought, is both simple and absurd.

The simple point is this: Mission Control sees light moving away from it at 186,000 miles per second. According to the last calculation, the astronaut also finds this value to be c, or 186,000 miles per second. No matter who looks at the beam, it always moves at the same velocity. Nothing could be simpler. The absurd part of the statement is this: the astronaut is chasing the light beam at a speed of 93,000 miles per second, yet the light beam still leaves him behind as if he weren't moving! What should a person believe? That which is simple but absurd or that which is obvious but untenable?

The character of science is to turn toward the simple, no matter how absurd, provided the theory is thoroughly consistent and provided there is firm experimental evidence to support the new idea. It was certainly thought absurd to suggest that the earth underfoot was whirling through space at breakneck speed. Yet today every schoolchild easily accepts this view, perhaps too easily.[7]

Is Light Velocity a Universal Constant?

It is.

I have introduced this constant in the invariant expression $(ct)^2 - r^2$ and then shown that light appears to move at the same speed no matter how hard an observer tries to catch up with it. One may claim, and rightly so, that this result was smuggled in at an earlier point when I claimed that $(ct)^2 - r^2$ is an invariant. More explicitly, if I claim that $(ct)^2 - r^2 = (ct')^2 - r'^2$ (where r' and t' are the different distance in space and the different time measured by an observer moving in space), I have already assumed that c is the same for all observers. It is necessary now to compare the consequences of this assumption, the original jump of the grasshopper, with observed facts.

The first fact is that light moves with the same velocity no matter how quickly the source moves that emits it. This evidence comes

[7] A few years ago, my daughter taught science to a high school class in a small town in Virginia. She got along fine until she started to talk about the moon landing. The class was highly amused. Did teacher really believe this TV hoax? I tried hard but unsuccessfully to provide my daughter with arguments that would convince her pupils. For them, the earth remained the beginning and end of reality.

from observations of double stars. In typical cases, these stars circle each other with a velocity as great as one ten-thousandth the velocity of light. Optical instruments, called spectrographs, can analyze the color of light. Because of phenomena that will be explained later in this chapter and the next, light from an approaching source is more violet-colored, while light from a receding source is more red. By using a spectrograph, one can see how the location of sharp lines superimposed on the bands of color (Fraunhofer lines) shifts back and forth as the two stars move around each other on elliptical orbits. The time of maximum approach of one star should coincide with the time of maximum recession of the other. In fact, the maximum shifts to the red and violet are seen simultaneously, just as one would predict on the basis of a constant speed of light.

If light were fired from the stars as bullets are fired from a gun or a ball from the hand of a pitcher,[8] the velocity of light would change. That this is not the case can be seen from the behavior of double stars that move in orbits around each other. If the speed of the emitter would be added to the speed of light, then the light from the approaching star would reach earth sooner than the light from the receding star. Since the distance of such double stars can easily be one-hundred light-years, the result would be that the signals would become tangled up during the period that they traveled. The signal from the approaching star would reach us almost four days too soon (compared to the case where light velocity is independent of speed of source), while the light from the receding star would be too late by a similar period. In fact, however, the light from double stars arrives in the same sequence in which it was emitted, giving a perfectly faithful and understandable picture of two stars circling each other according to Newton's laws. (See Appendix VI.) Thus the velocity of light cannot depend on the velocity of the emitting source.

One way to explain why this is true is to assume that a medium fills all space in which light is propagated. This was exactly what the physicists of the last century believed, and they introduced a quite incomprehensible substance, *ether*. They supposed that just as ocean waves are undulations of water and sound waves are pressure variations in air, so light waves are propagated through ether.

A "minor" difficulty with this theory was that ether seemed to elude any direct observation. If it were there, why did it offer no resistance to moving bodies? Did it penetrate all bodies without difficulty? If ether carried light by being displaced as light passed through it, it would have to resume its original position very rapidly

[8] In both these instances, the velocity of an object is affected by the velocity of the emitter.

in order to account for the enormous speed at which light travels. Ether would have to be either of exceedingly low density or much more rigid than air. Such a substance is a bit difficult to imagine. However, a scientist can imagine anything so long as it does not contradict the basic fabric of science.

A more subtle but serious worry was that the idea of ether contradicts Galileo's principle of the equivalent behavior of all uniformly moving bodies. Before Galileo, the earth provided the unmoving system, and all motion was considered in reference to the earth. That idea was abandoned when the earth itself was seen to be just one of many moving bodies. But if there were an omnipresent ether, it would have to be considered to be at rest. It could serve, as the earth had in the past, as the universal reference system with which all motion could be compared. If one were really at rest, one would notice it by not experiencing an "ether wind."

Galileo did not consider these problems, but he did attempt to measure the speed of light by opening and closing shuttered lanterns on adjoining mountains. The result seemed to be that light moved with infinite velocity or, at least, that light moved too fast to be measured by human reflex. If the speed of light were infinite, the strange result of the astronaut chasing the light flash would be explained. It would not be possible to chase it at half the speed of infinity. One would see distant stars instantaneously. The ether wind would not matter because light could not move faster than it was already moving—at infinite velocity.

Unfortunately, the speed of light turned out to be finite. It was first measured by Roemer, a Danish astronomer, and the results were presented to the Paris Academy in 1666. Roemer noticed that as the earth approaches Jupiter, the moons of Jupiter seemed to need less time to complete their orbit, and as the earth receded, more time was required. Of course, the time to complete their orbit was the same in both cases, but the earth gets the report sooner when it is closer to Jupiter. In this way, Roemer first noticed that light had a finite speed, and then he actually measured it. (Notice that the motion of the receiver has an effect on the time of reception; the motion of the emitter does not.)

After more than 200 years and a lot of other improvements, a more accurate and informative measurement was made by an American physicist, Albert Michelson. Michelson and a friend of his, Edward Morley, found that c equals 186,285 miles per second.[9] Thus light velocity is not only finite but, as compared to the motion of

[9] This figure has recently been revised to 186,282.396 miles per second.

heavenly bodies, is not even exceedingly fast. It is only about 10,000 times faster than the velocity of the earth in its orbit around the sun.

This gave Michelson the idea that it would be possible to measure the motion of the ether relative to the earth's movement. He could do this by conducting a race between light beamed through the ether wind at different angles. Upwind the light should move more slowly; downwind, more rapidly. Furthermore, since the earth reversed its velocity once every six months, one could not expect the ether wind to be zero at all times.

A direct measurement could not be made. That would have required a comparison of two clocks in different places. These clocks could not be compared except by looking at them using the very same light whose velocity the experiment was trying to measure. Michelson and Morley carried out an ingenious experiment in which only the findings of one instrument at one location were required.

Pythagoras Goes Swimming

These experiments can be compared with the situation of a man swimming in a river. Figure 2.4. shows that Pythagoras has rejoined the discussion, so I shall let him be the swimmer. Assume that he is a real athlete and can swim 3 kilometers an hour. The river flows at one kilometer an hour. Pythagoras is given the task of swimming one kilometer upstream and down and, when fully rested, swimming a distance of one kilometer across the stream and back. The question is: which can he do faster?

Upstream he will swim against the current, so his velocity will be 2 kilometers an hour. It will take him thirty minutes to reach a spot one kilometer upstream. On his return trip, he will move with the sum of the velocities, 4 kilometers an hour, so this part of the trip will take him only fifteen minutes, for a round trip time of forty-five minutes.

To calculate the time required for the one kilometer cross-river swim, one must use Pythagoras' theorem. An attempt to swim straight across the river would be thwarted by the current, which would carry the swimmer downstream. The swimmer's actual course in the water must compensate for this. The 3 kilometer speed must be divided between combatting the current and swimming to the spot directly opposite. Pythagoras, therefore, actually swims on a familiar diagonal, and one can compute this cross-river speed (v) by solving the equation:

$$3^2 = 1^2 + \mathbf{v}^2.$$

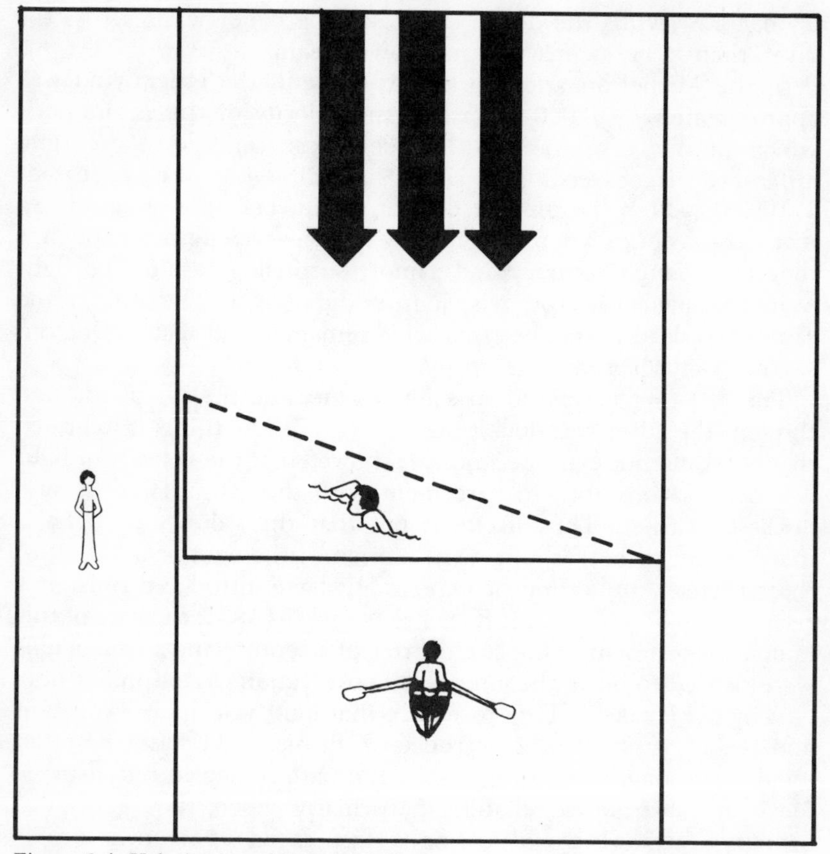

Figure 2.4. Velocity in a Moving Medium.

In order to visualize the situation using the figure, one must convert the actual velocity Pythagoras has in the water into the distance of his swim along the hypotenuse. In the 21.2 minutes required for his trip, the current will have flowed 21.2/60 kilometers, which means that the side of the triangle, a, is equal to 0.353 kilometers. The distance across the river, side b, is one kilometer. Therefore, side c, the hypotenuse, equals $1^2 + .353^2$, or c = 1.125 kilometers.

His speed (as assessed by an observer on the shore) is the square root of 8, or about 2.83 kilometers per hour. (In contrast, his speed in the water, as observed by himself or by a man in a rowboat resting on his oars, would remain 3 kilometers per hour.) He can swim across the river in 21.2 minutes:

$$\frac{60}{2.83} = 21.2).$$

Since the time remains the same regardless of the direction in which he crosses the river, his round trip will take only 42.4 minutes.

Pythagoras swims the distance in less time if he swims across the river than if he swims up- and downstream.

In the Michelson and Morley experiment, the ether wind was approximately 1/10,000 as fast as the velocity of the "swimmer," which in this case was light. Under these conditions, the time difference is exceedingly small, actually on the order of 1/100,000,000 of the time needed for the traversal. By using refined methods of optics where winning by a half-wavelength is enough,[10] they got enough accuracy and a most surprising result. The light beam that made the cross-wise journey did not win. The result of the race was a dead heat. The ether wind remained undetected. One can say more—its absence was proved.

The fact that it was not possible to measure the earth's motion through the ether was doubly surprising. For one thing, it indicated that ether did not exist. Secondly, it showed that the velocity of light is indeed a constant—in agreement with the "absurd" conclusion discussed earlier. Thus, it turns out that the velocity of light is independent of both the state of motion of the emitter and that of the receiver. Application of logic might have introduced Einstein's relativity and the invariant $(\mathbf{ct})^2 - \mathbf{r}^2$ as early as 1881, the date of this famous experiment. Almost a quarter of a century and a true genius were needed to make the jump from observation to comprehension.

Why did it take so long to realize that light velocity is a constant and the ideas about ether erroneous? Primarily because the ether wind was thought to be very small and hard to detect. But also, old ideas are difficult to change, particularly when they contradict obvious "fact." (I would define a fact as an apparently simple statement in which practically everyone believes.) Such "facts" were involved in two other ideas that were hard to change: the earth is at rest, and time appears the same to everyone.

Accepting the Absurd

When my son was not quite six years old, I told him a story about the moon, which I described as being very far away. I tried to give him some feeling of how far 239,000 miles really is. Then I told him that light travels very, very fast. It takes little more than one second to cover the distance between the earth and the moon. All this my young son accepted without too much astonishment and without any protest. Then I added the statement: light moves very fast, and nothing can move faster.

[10] Light beams can "interfere" with each other. (See Chapter 3, pp. 100-1.)

At that point I got a very polite, very justified query—why? I declined to answer. My son didn't pursue the subject, but the next day he asked, "Daddy, why can't anything travel faster than light?" I had achieved my objective of arousing his curiosity, and bit by bit I answered his question. It took all of six years. When my son was twelve and understood the main facts of relativity, I had high hopes that he would become a physicist. Instead, he chose philosophy. However, he is interested in the philosophy of science.

Presenting this material to a more mature audience, I shall contract the time of presentation, a practice appropriate to relativity. It has been suggested that it takes five years for a person to change his mind on a topic. When the topic is as basic as time, one should not be unduly bothered if the first presentation is hard to grasp.

Assume that I could travel with a much higher speed than that of light. Call that speed τ (the Greek letter *tau*) in honor of the hypothetical particle called tachion which is supposed to travel faster than light. (My conviction is that it doesn't exist.) If I could travel with the velocity τ, then I could be present at two events,[11] one of which occurs at $\mathbf{t} = 0$, $\mathbf{x} = 0$, and the other at $\mathbf{t}' = 1$, $\mathbf{x}' = \tau$. These are actually measurements made by an observer from his coordinate system, which is "at rest." I am moving with respect to him with the velocity τ. The observer, staying at Mission Control, will calculate the invariant as $(\mathbf{c} \times 1)^2 - \tau^2$. Since τ is greater than c, he will find that the invariant has a negative value.

From my perspective, I have been moving with the speed of τ from the first event to the second, so for me, both events occurred in the same place: $\mathbf{x} = 0$. I can say that the first event has occurred at $\mathbf{t} = 0$, and the second a little later. For me, the invariant will be $(\mathbf{c}\mathbf{t}')^2 - 0^2$. Obviously this calculation will give an invariant greater than zero, not negative at all. Hence, the invariants are not the same, and therefore $(\mathbf{c}\mathbf{t})^2 - \mathbf{r}^2$ is not an invariant. This is in conflict with Einstein's basic formulation. If one accepts Einstein's idea as correct, then one has found that nobody (alive or dead) can travel faster than c, the speed of light.

Repeating this important idea in a different way, assume—contrary to Einstein's ideas—that I could travel with an arbitrary velocity, with an almost infinite velocity. In this case, I could rush from one clock to another and compare their readings. I could verify in this way the time differences in systems moving arbitrarily with respect to one another. Then time could be considered absolute and

[11] For simplicity, the spatial distance in these two events will be considered only in the x direction of three-dimensional space. In this case, r, the distance between events, is the distance x from the origin.

invariant. Since Einstein's theory leads to different time and space intervals for different observers, Einstein also had to postulate that clocks in various coordinate systems cannot be compared readily. From this, it was clear to Einstein that he had to exclude the possibility of traveling at infinite speed.

Einstein's theory has its most important applications in two areas: in nuclear physics in the study of high energy particles[12] and in astronomy. First consider some further consequences of the theory that are illuminated in astronomy. We are living in the "suburbs" of a big assembly of stars, about 100 billion of them, called the Milky Way system or galaxy. It is possible to see, at a distance of about 2 million light-years,[13] another galaxy, called Andromeda, which due to its shape and appearance is called a spiral nebula.

Suppose I wanted to visit the Andromeda Nebula. Einstein tells me that nothing can move faster than light, and light takes about two million years to get to Andromeda. Since my physician tells me that I won't live that long, should I give up hope? The answer is no! I could still get there, at least in principle.

I shall need some engineers to perform feats beyond anything yet imaginable in order to make the trip possible, and I shall also have to violate some practical physiological limitations. If I can overcome these problems, I can then actually get to Andromeda in a rather short time, provided I count it in my "own" time.

To explain how it is that I can travel with a velocity less than c and still make the trip in a reasonable time according to Einstein's theory, I shall consider the invariant $(ct)^2 - r^2$ again. Assume my velocity is $0.999999999999c$, that is very high and exceedingly close to the velocity of light but available in principle. How close my assumed speed is to the velocity of light is demonstrated by considering a recognizable distance. Light almost reaches the moon in one second, and with this velocity, I would have fallen behind a light beam in getting to the moon by only fifteen thousandths of an inch (15 mils).

For an observer on earth, r is 2 million light-years, and t, the time required for the trip is somewhat more than 2 million years. In my expedition to Andromeda, the time of travel as seen by Mission Control will be

$$t = r/0.999999999999c,$$

and the $(ct)^2 - r^2$ can be calculated:

$$(cr/0.999999999999c)^2 - r^2$$

[12] This branch of physics did not even exist when Einstein proposed his theory of relativity. It has been the area where his theory has found most verifications.

[13] The measured value is 1.8 million light-years. To simplify the discussion I shall call it 2 million light-years.

which reduces to

$$r^2(1.000000000002 - 1)$$

or

$$(2 \cdot 10^{-12})r^2.$$

Since r, the distance to Andromeda, is 2 times 10^6 light-years away ($r^2 = 4 \times 10^{12}$), the final result is that the invariant calculated at Mission Control is eight light-years squared (that is, eight times one light-year times one light-year).

Now calculating the invariant as I—the traveler—see it, the trip starts behind the controls of my spaceship and ends there, so the distance, r, is zero. The invariant for me is

$$(ct)^2 - 0^2 = 8 \text{ light-years squared,}$$

and t, my time, is therefore equal to 2.83 years, which, considering the size of the undertaking, is a quite reasonable length of time.

How is this possible? The answer consistent with Einstein's relativity is that my watch runs more slowly, my heart beats more slowly (about fifty times a year according to Mission Control), and I age more slowly. This is called time dilation, first postulated by Hendrik Lorentz in 1904. Lorentz regarded his work as a purely mathematical device with no physical meaning. He failed to abandon his belief in absolute time (as well as his belief in ether) and did not notice the essential point that the meaning of time had been redefined. But his calculation was correct.

Philosophically, it is one thing to say that a watch runs more slowly and one's heart beats more slowly, and something else again to say that time itself has changed when observed from a different point of view. Actually, "my" point of view of the Andromeda trip is as well justified as anyone else's as long as I move with an unchanging velocity, that is, as long as neither my speed nor direction changes.

Of course, what has been described is not only oversimplified, it is exceedingly close to the velocity of light. I must first accelerate. Furthermore, when I arrive at Andromeda, I will probably want to visit, and, therefore, I shall have to decelerate. It is precisely in this acceleration and deceleration that the greatest engineering feats would have to be performed, and it is in these phases that I would pay attention to the stresses that my flesh and bones can bear. But these are points outside theoretical physics, and so can be neglected here.

Back within the realm of physics, the man at Mission Control will see me move away with almost the velocity of light, and, conversely, I will see him left behind at the same velocity. The same holds true for the rest of the universe, including Andromeda. On my trip, the old laws hold: nothing can move faster than light, and, indeed, nothing does. I shall spend only 2.83 years traveling. Therefore, it follows that I see the distance between the earth and Andromeda as only a little more than 2.83 light-years, whereas the man in Mission Control sees the distance as two million light-years.

The phenomenon just described holds not only for the distance to Andromeda, but for all distances measured along the direction of my velocity of motion. The mathematical formulation of relativity shows that the distances perpendicular to my motion do not change, but the parallel distances do. They are shortened—"contracted"—by a factor of nearly a million as the distance to Andromeda is. Therefore when I look back at the earth, it no longer appears an approximate sphere, but a disc about 8,000 miles across and only about sixty feet thick at its center.

The reader should beware of simple expressions like *as I see it in my rocket,* or *as Mission Control sees it.* Actually, what Mission Control and I see would be a blur. All either of us could do would be to make measurements. Even these measurements, if read straight from the apparatus, would have to be corrected in a fairly complex way for the appropriate time delays that the light took in transit. Therefore, these phrases really cover a number of operations and abstractions. All time signals would arrive so delayed that they would have to be measured very carefully and corrected on the basis of inferences. The simple expressions used in the text are really abstractions. Yet it is easier to think and speak in these abstractions whenever one considers things that move with similar velocity to one's own because they correspond in good approximation to easily perceived realities. Only when one considers very great velocities does the phrase *I see* lose its direct and simple meaning.

Going on with my imaginary adventure, I arrive at Andromeda, taking somewhat more than 2.83 years including acceleration and deceleration times. I have a look around, collect samples, make observations and measurements of scientific interest (including the investigation of life on solar systems in Andromeda) and then come home in less than three years. The trip has taken a total of about ten years of my life, but I can hope to get the Nobel Prize and a ticker-tape parade on Fifth Avenue, at least.

Unfortunately, on my return Fifth Avenue is no longer there. Four million years have passed on earth, humans have developed into creatures strange and horrible, especially so since they imagine themselves superior to me. They have, however, retained a curiosity about "antediluvian" beings, consider me interesting, and so house me in an air-conditioned zoo, complete with swimming pool, television, and, if I am very lucky, the works of Shakespeare.

From Fiction to Fact

These surprises are just some of the rather simple and somewhat sensational results of the theory of relativity. What I have described is not a fairy tale but an extension of real measurable effects. These effects cannot be measured on astronauts using the present state of technology, but they can be measured and checked for correctness on particles produced high above the earth in interstellar space or, more probably, by exploding stars. These particles are called cosmic rays.[14] The study of these particles and similar ones produced in laboratories is the modern field of high energy physics. This study occupies most physicists today and is the field where Einstein's theory has been most completely verified.

After cosmic rays strike the atoms of the upper atmosphere, an ephemeral particle, the mu-meson or muon, is eventually produced. In a short period of time (about $2.2 \bullet 10^{-6}$ seconds), the muon decays into an electron and two neutrinos.[15] The muon is so short-lived that even if it were traveling with the velocity of light, it would cover on the average only 660 meters in its life span. Muons are produced at an altitude that is about thirty times as great. Yet many of them are found at sea level.

How can such a short-lived particle pass through the entire atmosphere of the earth? The muon is behaving just as I did on my trip to Andromeda. From the perspective of the muon, the earth's atmosphere is so contracted that the muon reaches the earth in even less than $2.2 \bullet 10^{-6}$ seconds. From the perspective of the observers on earth, the muon "lives" much longer than .0000022 of a second. This is an actual example of time dilatation.

These results may be as shocking to some people today as the moving earth was to Copernicus' contemporaries, but they are just as real and eventually will be as commonly known. I do not know how many years were required for acceptance of the heliocentric theory by a majority of people. There was less formal education available at the time of Newton. The difficulty that so many people today have with

[14] Cosmic rays are nuclei of atoms moving at a speed almost equal to the speed of light.

[15] Neutrinos are peculiar particles which are very hard to detect and live forever. They exist all around us today and pass through us with the velocity of light. The current theory is that many of these neutrinos are generated near the center of the sun. (Recent observations indicate that these statements may have to be modified, but that does not change the argument given in the text.)

relativity may be part of the more general phenomenon that Snow pointed out: the mutual neglect of physical science by the humanists, and the humanities by those working in the "hard" branches of knowledge. This division at best represents a personal loss for each human; at worst it is a danger to a society that is dependent for its functioning on science and technology.

Yet I suspect that this explanation is not sufficient to describe the state of intellectual affairs today. The whole truth must also include the subdivision of knowledge to the point where one might say that knowledge has been atomized. Physics itself is subdivided into many areas, and similar statements can be made about mathematics. I suspect a similar trend in modern art. Styles are cultivated that are greatly appreciated by some but less appreciated by the average person. If intellectuals fail to awaken the awareness and cognizance of many people, they develop cliques. Individuals then lose interest in each other's work. Fewer and fewer people know what a certain development is about. The world may end up in a state where each person understands only what he or she is doing. Ultimately that person will wake up some night and wonder if even s/he understands what s/he is doing. When one loses sight of the common human experience, one may lose sight of reason itself.

This horrible state of affairs is the opposite of the pursuit of simplicity. For people who understand it, science does have the property of connecting phenomena that at first seem to be unconnected. In Einstein's theory, the main point is that instead of talking about space and time separately, one needs to talk about events taking place in space-time. The relationships can be grasped only if space and time are simultaneously taken into account. The next section will introduce a similar phenomenon.

Unifying Momentum and Energy

There is another instance in which Einstein's theory produced a simple relationship between apparently different laws of physics. The invariant $(\mathbf{ct})^2 - \mathbf{r}^2$ is not the only anchor in the theory of relativity. Just as this invariant establishes the relationship between time and space, Einstein provided a further anchor for the relationship between two other quantities: energy and momentum.

The energy being considered here is the energy connected with the motion of an object, commonly called kinetic energy (K.E.). This energy is defined as the mass ($\mathbf{m}$) of an object times the square of its velocity ($\mathbf{v}$) divided by 2:

$$\mathbf{K.E.} = 1/2\ \mathbf{mv}^2.$$

Webster's conventional dictionary says that mass is one of the fundamental quantities on which physical measurements are based, the other two being length and time. (The reader may now have a suspicion of what is coming.)

Velocity, like space, has three components,

$$v_x, \quad v_y, \quad v_z,$$

which represent the velocity in the **X**, **Y**, and **Z** directions. The total velocity, **v**, is connected with these in the same way that **r** is connected with **x**, **y**, and **z**, that is,

$$v^2 = v_x^2 + v_y^2 + v_z^2.$$

Because kinetic energy depends on the state of motion, it will appear different to different observers.

The quantity that is paired with energy (E), as space is paired with time, is momentum (p). The momentum, in turn, is mass times velocity, or

$$p = mv,$$

so it also differs depending on the observer.

Einstein constructed an invariant that unlike energy and momentum remains the same for all observers. This second invariant of Einstein is

$$E^2 - c^2p^2.$$

The remarkable thing about this invariant is that it requires a redefinition of energy in order to be valid. If one identifies energy as kinetic energy (**K.E.** = 1/2 mv^2) and momentum as **p** = **mv**, the resulting expression is not an invariant.

Einstein noticed, however, that an invariant can be produced by giving a new and surprising interpretation to energy. Einstein introduced his famous connection between energy and mass,

$$E = mc^2,$$

energy equals mass times the square of the velocity of light. Since the velocity of light is so large, this gives an enormous value for **E**. Then in order to insure that the invariant really does not change whether one is in Mission Control or in the rocket, Einstein assumed that the mass itself changes when the velocity changes. The term refers to a "rest" mass in the special case where v = 0, and Einstein redefined mass as:

$$m^2 = \frac{m_o^2}{1 - v^2/c^2}$$

The description of energy then becomes: $E^2 = \dfrac{m_o^2c^4}{1 - v^2/c^2}$

This becomes clearer if one considers the invariant $E^2 - c^2p^2$. Remember that $p = mv$, so the invariant can now be written

$$E^2 - c^2(mv)^2 = \frac{m_o^2c^4 - c^2m_o^2v^2}{1 - v^2/c^2}$$

which reduces to

$$E^2 - c^2p^2 = (m_oc^2)^2.$$

This seemingly miraculous simplicity came about because of the tricky way in which Einstein introduced $1 - v^2/c^2$ into the denominator.

The point is that if one travels along with the object—that is, has the same coordinate system as the object whose energy is being considered—one has a velocity equal to zero. Energy equals mass times the velocity of light squared. Therefore, m_o can be called the rest mass, that is the mass characteristic of an object at rest, and m_oc^2 is then the corresponding rest energy.

How does all this relate to the concept of kinetic energy from which I started? Einstein's formula for energy can be approximately written (using a little mathematics which laypeople may take on faith) as

$$E = m_oc^2 + 1/2\, m_ov^2 + 3/8 m_ov^4/c^2 + \ldots$$

where the three dots mean that more terms should follow. All these terms are negligible for small values of velocity.

This formula contains more than kinetic energy. First, one can say that for $v = 0$, E becomes the famous m_oc^2 as has been shown. Second, the kinetic energy, the additional energy of an object which is due to its motion is therefore the value of energy in excess of m_oc^2. In this way one can describe kinetic energy as

$$K.E. = 1/2 m_ov^2 + 3/8 m_ov^4/c^2 + \ldots$$

In the case of low velocities where v/c is quite small, the second term is truly negligible, and one returns to the usual value of kinetic energy which is $K.E. = 1/2\, m_ov^2$. Thus Newtonian physics was modified while remaining quite usable. Actually it took the great amounts of energy concentrated on small particles that occur in nuclear physics to give a quantitative proof of $E = mc^2$. These experiments did not come along for quite a few years after Einstein suggested his new relationship between energy and mass.

It used to be said that energy is conserved and so are the three components of momentum. The expression $E^2 - c^2p^2$ could not remain an invariant if one of these conservation laws held and the others did not. In mathematical language, it is said that there is only a single conservation law: the energy-momentum vector is conserved. So now, previously separate laws have been connected.[16]

The theory of electricity and magnetism—the last one to develop before relativity—was much less affected by Einstein's theory. It is well to remember that as simple a thing as the concept of kinetic energy did require modification to be compatible with the ideas of relativity. The equations of electromagnetism developed by James C. Maxwell in the second half of the 19th century did not need any changes at all. However, the connection between the concepts of electric and magnetic fields did.

Even before Maxwell's time, it was clear that an electric charge produces an electric field but no magnetic field as long as the charge is at rest (static). A moving electric charge (current) produces both an electric and a magnetic field. The idea is close at hand that any electric field viewed by someone in motion will appear as an electric and a magnetic field because for this observer the electric charge that produced the electric field appears to be moving.

The converse point—that a magnetic field without the presence of an electric field will appear as a magnetic field plus an electric field to a moving observer—was not clear before Einstein. The remarkable point is that Einstein did not need to change Maxwell's equations. He only had to clarify the way in which magnetic and electric fields transform into each other if one moves relative to them. Electromagnetic laws were simplified in Einstein's scheme; specifically, Maxwell's equations remained equally valid regardless of the state of motion of the observer.

By these means, broad regions of physics have been reduced to a simpler form. But this is not the end of Einstein's accomplishments. What was in some ways the most remarkable achievement was still to come. It took Einstein a dozen more years of hard work.

When One Is Low, One Is Slow

I have considered the idea that various forms of uniform motion are equivalent and have specified the logical conclusions stemming from this. Einstein noticed that there was another law of equivalence

[16] A vector is commonly represented by a directed line segment whose length gives its magnitude but which also has a direction in space. Velocity, acceleration and force are other examples of vectors. In relativity, a vector also has an orientation in time. Its magnitude is the square root of the invariant. This law means, as one application, that when two particles collide, the sum of their energy-momentum vectors will be the same before and after collision. Various observers will assign various values to the energy and momentum before and after, but each will agree that neither the sum of energies nor the sum of the momentum values nor even the sum of any component of momentum will have changed.

in physics—that all bodies are accelerated in precisely the same way by the force of gravitation. One consequence of this equivalence would be that if one were in a freely falling elevator, one couldn't tell that there was any motion or gravitation. During the fall, the elevator and the bodies in it fall in the same way. One might as well be out in free space. In the end, of course, one would notice—at the bottom.

Similarly, an astronaut circling the earth does not notice that he is accelerated once his orbiting speed has been reached. A pencil held up and then released does not drop because it remains in the same orbit as the astronaut. Long before astronauts, Einstein had the idea that relativity could be extended to accelerated systems as long as the acceleration is due to gravity. The most surprising result of this investigation emerged after a rather complicated argument, and that result was a systematic explanation of the phenomenon of gravitation, including novel phenomena which show up in extremely strong gravitational fields.

The basic observation that Einstein made was that people cannot tell whether they are being accelerated or whether they are being acted upon by a gravitational force. This "principle of equivalence" can best be visualized by considering a closed box to which an angel has attached a rope and is pulling "upward." (This archaic source of propulsion can, of course, be replaced by a rocket engine below the box.) Inside the box, there is pressure on one's soles, but whether the pressure is due to gravitational force or whether the angel-engine is accelerating the box can't be determined from inside.

According to this idea—that gravitation is equivalent to acceleration—Einstein drew up equations that were valid in any coordinate system. He came to an amazing number of results, all of which have turned out to be correct. One of the most interesting can be connected with the trip to Andromeda.

It is reasonable to ask in the spirit of relativity whether during this trip I moved relative to the earth or the earth moved relative to me. Recall that my round trip to Andromeda required some ten years of "my" time, but four million years passed on earth. Why can't one claim that it is the other way around—if the man in Mission Control finds that my heart beats only 50 times a year (his year), why can't I assert that his heart beats only 50 times a year (my year)? Indeed, I am justified in saying this while I am in uniform flight. By the time I reach the vicinity of Andromeda, I conclude that the man in Mission Control aged a little more than two minutes.[17] How can it be that upon my return I find that four million years have passed on earth?

[17] The compression of time seen in considering the time I required to reach Andromeda has an equivalent in the compression of time here.

This riddle is answered by the fact that my system is distinguishable from that of Mission Control because of the accelerations that I experienced. These accelerations are detectable. Both I and Mission Control will agree that I was accelerated and not the earth. There were measurable effects to which I was subjected that the people on earth did not experience. This is the source of the difference which can be noticed inside my rocket.

According to Einstein's principle of equivalence, my accelerations and decelerations are equivalent to very large gravitational fields, so large that if I wouldn't arbitrarily have chosen to forget about the facts of physiology, I would have been crushed. However, two of these occasions of change in motion give rise to no trouble with time. During the take-off from earth (when I am accelerated) or landing on earth (when I am decelerated), watches can be compared directly. Not so when I arrive at Andromeda and have to slow down, nor when I speed up again on my return journey to earth. In both these cases, I am two million light-years from home, and watches cannot be compared in any direct manner. The lack of similarity of the observations of Mission Control and my own must be blamed on the deceleration and acceleration I experience on Andromeda. According to Einstein's gravitational principle of equivalence, I will feel during this acceleration (and deceleration) as if there were an enormous gravitating body located beyond Andromeda. Indeed, if my head is pointing toward the earth, there will be great pressure on my soles, just as if I were at rest with a big attracting mass below me.

Einstein's idea is that huge time differences of four million years must have occurred from my point of view during these two periods of deceleration and acceleration at a distant place. They could not have occurred on any other occasion. This means that when one is close to a gravitating body, time progresses more slowly than where there is neither gravity nor change in speed. The lower I am in this field of gravity, the slower time passes. This effect is known as Einstein's gravitational red shift, because the observed radiation from the lower vibration of electrons in atoms near a gravitating body gives rise to an observed shift of spectral lines[18] toward the red.

By very, very careful measurement, a difference in passage of time can be detected between the top and bottom of the not overly impressive Harvard Tower. The vibrations at the top of the Tower compared with the vibrations at the bottom are shifted by an incredibly small amount which is nevertheless detectable if you are an excellent Harvard physicist with a predilection for the minute and

[18] Spectral refers to the spectrum of light. Depending upon the length of the waves of visible light, white light is separated into its component colors when the light is passed through a prism. Red light has the longest wavelength of visible light; therefore, it vibrates more slowly. (See Figure 3.8.)

exact. In this case, of course, the source of gravitation is the earth itself. It is a remarkable theory which recognizes the same phenomena whether they are connected with the exploits of astronauts (greatly exaggerated in this discussion) or with the exceedingly refined measurements of atoms in experiments carried out at home. To my mind, this introduces simplicity in a way least expected.

The Uses of Curved Space

The story of the red shift is only a part of the brilliant theoretical structure which is best called Einstein's theory of gravitation.[19] One of its many consequences is the derivation of Newton's inverse square law of gravity. (See Appendix VII.)

Newton arrived at this law empirically, using Kepler's description of planetary motions. He was obliged to answer the question of why it should be so with the now famous

> I frame no hypotheses; for whatever is not deduced from phenomena is to be called a hypothesis; and hypotheses, whether metaphysical or physical, whether of occult qualities or mechanical, have no place in experimental philosophy.

Despite these brave words, Newton not only was deeply interested in metaphysics, he put a lot of effort into what most people today call alchemy. But just as Newton was able to derive Kepler's laws from a simpler basis, Einstein was able to derive Newton's law of gravitation from a more fundamental and hence simpler standpoint.

Einstein's approach was to frame the problem in a four-dimensional curved space-time, and with this concept he was able to show that matter "warps" space and produces a curvature in the vicinity of its location. This warping replaces the classical gravitational field. One may visualize this by imagining a curved surface out in space far removed from any gravitating object. Also imagine that an object moves in such a manner that it cannot leave the curved surface, but that otherwise no force acts on it. Then the path of this object will be bent because the surface is curved, and this bending will be actually observable if the two-dimensional observer cannot see anything outside the surface. In this way one acquires a two-dimensional picture of a motion which in Einstein's theory requires added dimensions.

To describe this gravitational theory in curved four-dimensional space requires more mathematics than can be explained in a few

[19] Many people call it Einstein's general theory of relativity.

paragraphs. But I can go so far as to explain how Einstein used the word *curvature.* This can be done by using a two-dimensional example again, such as the surface of the earth. Imagine what happens to an arrow when one carries it on a closed path, trying always to point the arrow in the same direction. If this is done on a plane surface, the arrow will retain its original direction when it returns to its point of departure after a complicated journey. This is not true in the case of a curved surface, such as that of the earth.

Imagine an arrow located at the north pole and pointing towards New York. Carry the arrow south through New York to the equator. Now carry the arrow west, still pointing in a southerly direction. After arriving at the longitude of Hawaii, carry the arrow north, while continuing to point the arrow south. In the end, back at the north pole where the trip started, the arrow points towards Hawaii and not at New York. Even though the arrow was carefully pointed south throughout the trip, it ends up pointing in a new direction when one arrives at the original starting place. This is a characteristic of curved surfaces which is used to define them. Einstein adopted this definition (which had been introduced into mathematics a century earlier by Gauss and Riemann). Then Einstein added an essential hypothesis: that space curvature was due to the mass found in space.

Many surprising predictions can be made using the concept of curved space—like the bending of light rays in the vicinity of massive stars. This latter has been verified by observing the stars just outside the solar disc during total eclipses of the sun. The simple reason is that the light beam in passing the sun "falls" toward the sun. To get a quantitatively correct answer describing the path of the star's light, one must calculate the influence of the strong gravitation near the sun. This is a truly remarkable verification of an even more remarkable prediction.

A recent astronomical entity that has created a lot of publicity fits perfectly into Einstein's theory, even though he did not mention it. It is the "black hole." It is known that objects can't escape from the earth unless they attain a velocity of some 7 miles per second or more. This is called the escape velocity; astronauts must reach it before they can take off for the moon. On a body much more massive and dense than the earth, the escape velocity may reach or exceed the speed of light. In this situation, not even light is able to escape. If there are such objects in the universe, and there is reason to believe that they exist, then they have the very peculiar property that things may fall into them, but they can never get out again!

Resistance to Unification

When Einstein completed his work on general relativity, he was not yet forty years old. He planned to do much more. Having reduced time, space and gravitation to a geometric explanation, he hoped to accomplish the same thing in all of physics. At that time, in the second decade of this century, there was some basis for the hope that if electromagnetic theory could be included in a broad geometric approach, all of physics would be explained and simplified to the one basic science that the Greeks had considered the center of exact thinking, geometry. For the rest of Einstein's life, his main purpose was this unification, this tremendous simplification. He did not succeed.

It was not for lack of trying. For almost forty years, with more and more elaborate mathematics, Einstein pursued his unified field theory. Shortly before his death, he published an attempt which, at least in mathematical details, was anything but simple. I heard about a reporter who asked him at that time whether or not he had succeeded, whether or not his theories were correct. Einstein's answer was: "I do not know; come back and ask me in twenty years."

Of all the people who have sought simplicity, Einstein was probably the greatest and the most successful; even he did not succeed fully. He did not complete the pursuit of simplicity. He had to leave the continuation of that pursuit to the future. Additional facts about the physical world have been discovered both before and after his death in 1955. These new facts are not in conflict with Einstein's laws, but they disclose a new world full of previously unsuspected complications. By having simplified what is known, physicists have been led into realms which as yet are anything but simple. That at some time, they, too, will appear as simple consequences of a theory of which no one has yet dreamed is not a statement of fact.

It is a statement of faith.

A NOTE ON CALCULATING THE CHANGES IN TIME AND DISTANCE BETWEEN TWO EVENTS VIEWED FROM SYSTEMS OF DIFFERING RELATIVE VELOCITY.

According to conventional ideas, distances will change when viewed from two different systems, one "at rest" and the other system moving relative to it with velocity v along the x direction. (For greater simplicity, I will consider motion and direction only along this one direction, replacing r by x.)

The change in distance under the old system can be written as:

$$x' = x - vt,$$

where x represents the distance in the first system, x′ the distance as viewed from the second system, and vt (the velocity multiplied by the time) is due to the shift of the coordinate system from the origin in the time t. If, in keeping with the discussion of the invariant $(ct)^2 - r^2$, one wants to use the quantity (ct) instead of t,

$$x' = x - v/c\ (ct).$$

Similarly, conventional ideas represent $t = t'$, or $ct = ct'$.

The new transformations are more symmetric and also contain a square root expression in order to insure that $(ct)^2 - r^2 = (ct')^2 - r'^2$. The transformations are

$$x' = \frac{x - \frac{v}{c}(ct)}{\sqrt{1 - \frac{v^2}{c^2}}} \quad \text{and} \quad ct' = \frac{ct - \frac{v}{c}(x)}{\sqrt{1 - \frac{v^2}{c^2}}}$$

This, as can be shown by simple mathematical operations, preserves the invariant. It is only a little more difficult to show that these equations for x′ and ct′ are the only ones which conserve the invariant and at low velocities reduce to the conventional equations.

The remarkable fact is that the new equations were derived before Einstein's work by a Dutch physicist, Hendrik Lorentz, and so the relationship between the new and old coordinates are called the Lorentz transformations. Lorentz derived these formulas in a purely formal manner to explain the Michelson-Morley experiment on the ether wind. Einstein's phenomenal contribution was to recognize that x′ and t′ are the only values for distance and time differences that make sense from the point of view of the moving observer and that absolute motion and phenomena like the ether wind do not exist.

Chapter Three

THE SCIENCE OF PARADOXES

The concepts considered in this chapter may be the strangest of the strange concepts developed in science thus far. As with relativity, the point is not that the subjects are complicated but rather that these ideas are both unexpected and contrary to commonplace notions. At the same time this part of science makes it possible to explain the structure of the atom and of matter in general in a remarkably complete manner.

Knowledge of the atom today seems as reliable as knowledge of the solar system was when the mathematical alchemy of Newton transmuted natural philosophy into physics. It is surprising that some scientists, among them no lesser genius than Einstein, have found this theory difficult to accept in its entirety. The theory has proved remarkably successful, and Einstein's opposition was not based on evidence.

People often complain that scientists disagree on important public issues. Usually the explanation is that some scientists make statements outside the field of their competence. In the controversy between Einstein and Bohr (the originator of atomic theory), one can see an example of a deep disagreement of a truly intellectual nature between the two most creative men of the century. And yet, their only aim was the search for true simplicity.

The Atoms of Democritos

The rather amazing proposal that atoms exist was made more than 2000 years ago by Democritos in Thessaly. A legend has it that Democritos had been invited to give a "seminar" in Plato's academy.[1] After listening to the presentation, Plato sent Democritos to Hippocrates for treatment: it was obvious that Democritos had lost his mind. However, after Hippocrates applied the ancient equivalent of psychoanalysis, he appeared arm-in-arm with his patient, saying, "If this man is crazy, so am I."

Hippocrates was exceptional. In general the idea of atoms was rejected by the Greeks and almost everybody else until about 200 years ago when modern chemistry began. One should realize that while the reasons for believing in atoms were quite persuasive, there was a very solid reason for not believing in them. Why should matter have a limit beyond which it cannot be subdivided? There is no similar limit in direct experience. Why assume such a unique and peculiar idea? What difficulties would spring from such a concept? (The answer to the last question is: plenty.)

One reason for accepting the atomic hypothesis is that with a few simple (though unusual) statements, one can explain a lot of phenomena. For example, if water freezes, it changes its properties completely. It even looks like a different substance. If it is heated sufficiently, it melts and becomes water again. If it is boiled, it becomes vapor, and in a superficial way one can say that it has disappeared. However if water vapor cools, droplets form, and one gets the water back.

Using the atomic hypothesis, these changes are explained today by saying that water consists of molecules, and the difference between water, ice and steam is simply the arrangement of these molecules. In water, they are closely packed but disorderly. In ice, they form a crystal that is an orderly lattice. In vapor, they are widely separated and move almost freely. The ancient Greeks observed the phenomena of freezing and melting, vaporization and condensation. Democritos made these radical changes the basis of his hypothesis.

He believed that everything consisted of very, very tiny units that he called *atoms* (the Greek word for *indivisible*). He thought that the reversible changes in the properties of a material could be explained as changes in the ordering of these smallest pieces. Democritos imagined that there was an almost limitless variety of atoms, so there is one basic difference between his understanding and ours. His wide

[1] I heard this legend in Greece in an atomic energy laboratory—named Democritos.

variety of "smallest pieces" are now known as *molecules* rather than atoms.

Molecules and Crystals: A Giant Step Beyond the Microscope

About 1800, as chemistry started to emerge, scientists discovered that matter consists of molecules made from only some 90 different kinds of atoms. (Chemical substances consisting of one kind of atom are called *elements.*) These atoms are "indestructible," and they can enter into various combinations. However the atoms of one element differ intrinsically from atoms of other elements and from the molecules they form. For example, the hydrogen and oxygen atoms differ thoroughly from each other and from the water molecule they form. By contrast, water vapor differs from ice only in the arrangement of the molecules.

What happened in the 19th century was that chemical formulas and molecular structures were worked out for an incredible range of substances. Literally hundreds of thousands of compounds were deciphered and understood as arrangements of the constituent atoms, and then these and similar molecules could be synthesized. Both substances which occur in nature and substances which had never been seen before were obtained. Thereby chemistry immensely simplified the description of the structure of matter. All substances could be reduced to arrangements of these "immutable" elements. But this was not all. There were a number of additional fine and amusing points.

In molecules, atoms occupy defined positions. In the water molecule, one oxygen atom and two hydrogen atoms are attached in a rather peculiar fashion. The hydrogens are the same distance from the oxygen, but the three atoms don't lie on a straight line. (See Figure 3.1.) Carbon and oxygen atoms, in the case of carbon dioxide, have a different pattern: there the three atoms do lie on a straight line: **O—C—O.**

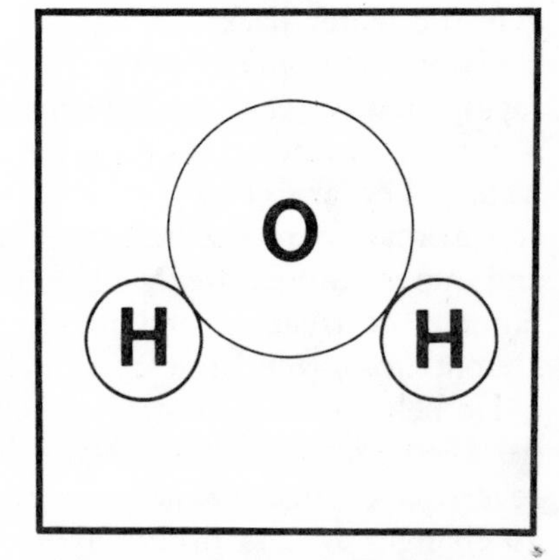

Figure 3.1. The Water Molecule.

In crystals, molecules occupy regular positions; the result of this regular positioning can be seen with the naked eye. When one observes ice crystals formed under conditions where they grow slowly—as in the case of snowflakes—one sees a remarkable six- or eight-fold symmetry, rather leaf-like in appearance. (See Figure 3.2.) Snowflakes under the microscope are indeed quite beautiful; merely looking at such a picture makes it plausible to think of an underlying orderly arrangement.

Figure 3.2. Snowflakes Seen Through An Optical Microscope.

If one goes into a salt mine (where salt crystals have grown slowly), one sees that the faces of the crystals are mutually perpendicular, as in a cube, corresponding to the cubic lattice structure. Actually common table salt, sodium chloride, has equal numbers of sodium and chlorine atoms and forms one of the crystal structures most easily visualized. (See Figure 3.3.)

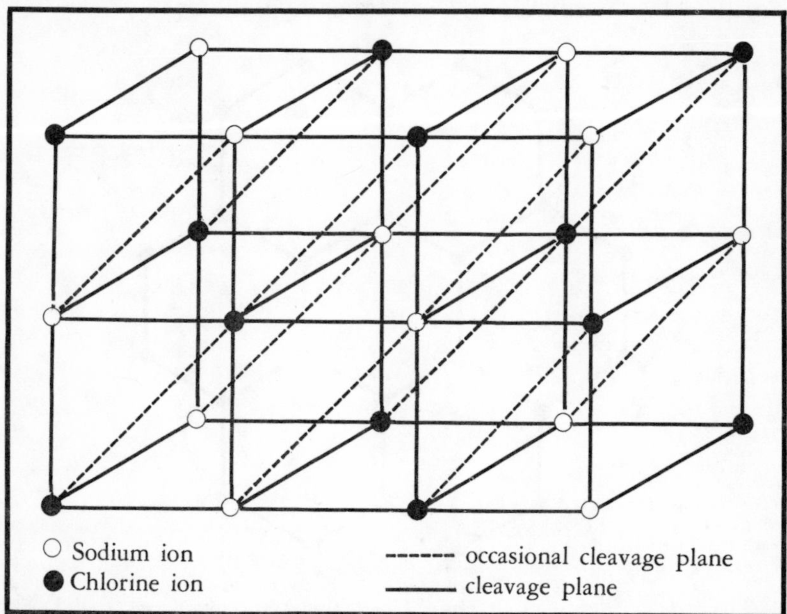

Figure 3.3. The Crystal Structure of Table Salt.

Crystals can be split in a variety of ways but not arbitrarily. They have a number of well defined angles formed by the natural cleavage planes. These planes are the ones on which atoms lie closely spaced on the same surface. The salt crystal will split most easily along the planes indicated in the figure. These planes are perpendicular to each other so that the microscopic structure of the cubes manifests itself to the eye in the macroscopic appearance of the crystals that one sees. In general, there are several different planes along which one can split a crystal. The fact that many atoms must be in a plane imposes relations between the cleavage angles of a crystal. The planes thus predicted are fully verified by experience.

Graphite has a honeycomb-like structure. It consists of sheets of carbon atoms bound tightly to one another and occupying the corners of hexagons. (See Figure 3.4.) These sheets are then loosely connected as they lie over one another. The result is that graphite comes apart easily, flakes, a phenomenon illustrated when one writes with a "lead" pencil. Graphite is also useful as a lubricant in machinery. In this application, graphite in exceedingly thin sheets is used. All these examples, water, salt and graphite, demonstrate both that the crystalline structure of matter is closely related to a material's macroscopic properties and that the crystalline structure itself is explained by the atomic and molecular hypotheses.

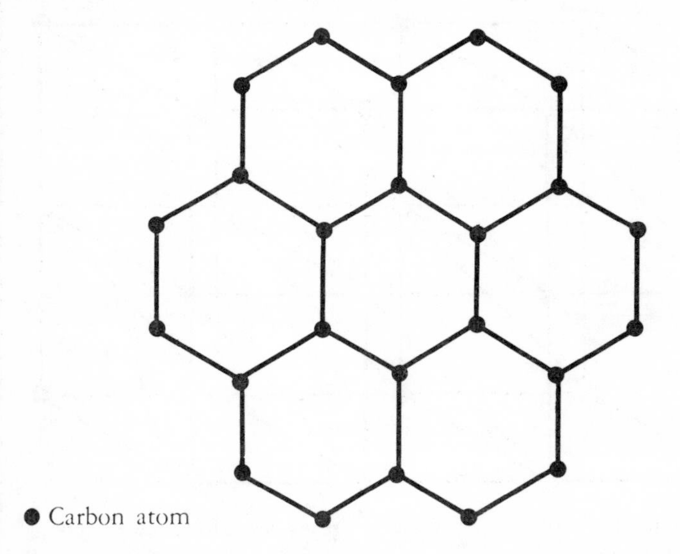

Figure 3.4. The Structure of Graphite. This is the arrangement seen when looking down on a layer of graphite.

The Chemistry of Electricity

Michael Faraday, a book-binder turned chemist, outdid both physicists and mathematicians by establishing some of the fundamental ideas of electromagnetism. He was the man who introduced the idea of "lines of force" in magnetism and electricity, the idea I used earlier to illustrate gravitational attraction. Faraday was convinced that the nature of atoms was electrical and magnetic, and he did a lot to prove this point.

One important finding was that dissolving salt in water doesn't lead to sodium and chlorine atoms "swimming" around. Rather one finds positively charged sodium ions and negatively charged chlorine ions.[2] If an electric current is sent through salt water, the ions will move to the positive and negative terminals of the current—sodium ions to the negative terminal and chlorine ions to the positive. These two different materials are then deposited on the terminals (if one disregards the fact that sodium immediately reacts with water). One finds that a certain amount of deposited matter is associated with a particular amount of electricity. There is a fixed ratio for each particular ion between the amount of electricity given off by the terminal and the amount of material deposited on the terminal.

As I have mentioned, sodium chloride consists of equal numbers of sodium and chlorine atoms. This is one particular case of a general rule: atoms combine to form molecules in fixed ratios. If one tries to add too much of one element or another, the excess is simply left over.[3]

Chemists already knew of the regular proportions of atoms and their relative weights in forming molecules at Faraday's time. However, Faraday added a further statement that atoms or groups of atoms combine with electricity and thereby form ions. In addition, a particular atom will combine with a particular fixed amount of electricity. It seems that electricity consists of smallest parts which are called electrons. It is this quantity of electricity which determines the formation of ions and explains the currents that flow in salt solutions. The electrons also proved fundamental in explaining the formation of molecules.

To connect these ideas with the sodium chloride crystal, I must point out that this crystal consists of sodium ions and chlorine ions.

[2] *Ion* comes from the Greek *ienai* meaning *to go*, so ions are "walking" atoms. They walk in definite directions, provided there is an electrical field to motivate them.

[3] This rule holds true in the most simple cases. Sometimes the same kind of atoms can combine in more than one way. For instance a carbon atom can combine with one or, more commonly, with two oxygen atoms.

The Greek word is now less descriptive because in a crystal, the ions do not wander about. Each positive sodium ion is surrounded by six negatively charged chlorine ions and vice versa. The simple arrangement is due primarily to the fact that opposite electrical charges attract.

Statistics and Temperature

One of the most interesting developments of the atomic theory was that it connected heat and the idea of moving atoms. As early as the 17th century, Francis Bacon, Robert Boyle, Robert Hooke and Isaac Newton all had clearly affirmed the correct nature of heat. Hooke stated it thus: "heat is a property of a body arising from the motion or agitation of its parts." An older opposite view, called the caloric theory, held that heat was an "element," a type of "fundamental fluid possessed by all bodies." The caloric theory not only survived these four scientists' attacks but gained common acceptance and held it for over one hundred years. The experiments of Benjamin Thompson, later Count Rumford,[4] finally killed the caloric theory of heat, but the details were not worked out quantitatively until quite late in the 19th century. The understanding of heat energy as the motion of molecules helped to strengthen the idea of conservation of energy and that led to interesting and important contradictions which I shall discuss later.

There is a theoretical temperature called absolute zero where all atomic motion ceases. It is as cold as it can get (and that is quite a bit colder than the winter temperature in Chicago, an amazing fact to those who have lived there). At ordinary temperatures, the heat in a room is the motion of the nitrogen and oxygen molecules of air (plus a few others), and this motion exerts a pressure on the walls. At this point a quite remarkable approach to heat phenomena started, and this approach was the most important consequence of the theory of heat. It is the use of a statistical theory rather than a straightforward progression from individual cause to individual effect.

What is the meaning of temperature? At a given temperature, the motion of the molecules gives the material a certain heat energy. The

[4] The world of scientists was then as now connected by social as well as professional ties. Consider this coincidence. Thompson is credited with destroying the "elemental-caloric theory of heat." Lavoisier is credited with destroying the "elemental-phlogiston theory of burning" (according to which burning consisted of losing phlogiston rather than acquiring oxygen). Lavoisier was born a French nobleman and lost his head in the French Revolution. Thompson was born in America, fled as a Tory during the American Revolution, and finally secured his long-sought nobility in England. He then married Lavoisier's widow. It seems that Countess Lavoisier-Rumford had a predilection for the noble slayers of the mythical monsters in science.

oxygen and nitrogen molecules of a room share this energy among themselves. They move in different directions and with different speeds, but there is an average energy which is the same for nitrogen molecules and oxygen molecules. This average energy of motion depends on the temperature and in turn is the measure of temperature.

The science of heat has become exact and intricate. It may come as something of a surprise that "exact" laws can be derived from probabilities or statistical situations. Statistical laws will give more and more exact results as the aggregates get larger. When a person throws the dice, even and odd should occur equally often; the ratio of even to odd approaches one when the number of throws is very large. Similarly, because of the immense numbers of molecules in even a microscopic piece of matter, the properties of matter that depend on the aggregate behavior of molecules are very sharply defined.

One example of the theory of heat is the theory of melting. The reproducible exactness of the transition of a material from one "phase" to another, such as the transition of ice to water, is truly remarkable. Why do these transitions take place at sharply defined temperatures (called melting points)? Why shouldn't an object start getting soft and its parts runny well before it melts?[5]

The qualitative reason is not too hard to understand. Begin by comparing a crystal with a fluid. The crystal is orderly—the fluid disorderly. Packing a suitcase by randomly throwing things into it will not result in an orderly arrangement. There are just too many ways for a disordered arrangement to occur for order to be produced by chance. Order is an arrangement that is inherently improbable.

Another improbable situation is that excessive energy will be given to a single molecule. This reduces the number of ways in which energy could be distributed among molecules. Thus, while the energy of each individual molecule will, in general, be slightly different, it is very improbable that more than a small portion of molecules will be appreciably different in energy from the majority of the aggregate.

The energy in the crystalline state is lower. When a crystal melts, the disorder increases. An increase in disorder is an increase in the probability of occurrence. At the same time when melting occurs, the energy increases, thereby decreasing the probability of a liquid state. Melting occurs at the exact temperature where the improbability of order in the crystal (connected with low energy) equals the improbability of excess energy in the liquid (connected with disorder).

[5] Actually, the transition is gradual for disorderly (or amorphous) substances, like glass. Only for orderly crystals is the melting point sharp.

The definiteness of the melting temperature is due to the competition between two very big and sharply defined probability influences. The melting point of a tiny crystal consisting of only a hundred atoms would not be very definite. In this case, one might argue that such a small piece of matter is not a proper crystal anyway because the high degree of order associated with a crystal requires the cooperation of an extremely large number of atoms.

It is obvious that the atomic hypothesis explains more than the chemist's representation of molecules as combinations of atoms. The atomic hypothesis makes it possible to understand crystals, transitions between states of matter, heat and temperature, and electric properties of matter. Characteristics of atoms may even have something to do with the properties of life itself—but this no one knows.

Measuring the Atom: Many Methods, One Result

Despite these successes, there was still the question: *do atoms really exist?* By the end of the 19th century, so much was explained by the atomic concept that it was sheer heresy to doubt it. However, a German physical chemist, Wilhelm Ostwald, and an Austrian physicist, Ernst Mach, managed not only to doubt it but also to attack it. Their point was the old one made by Plato: it's absurd to believe that matter can't be subdivided indefinitely. The success of the atomic theory from their point of view was based on the fact that matter behaves *as though* it consisted of atoms. They thought that if one wanted to believe in atoms, then one should consider those atoms infinitely small.

Their criticism was important because it stimulated attempts to prove conclusively that atoms exist. The proof was ultimately provided by measuring the size of atoms. The measurements were made in a number of different ways, all of which yielded the same size for the atom. (I will soon introduce the great paradoxes due to the existence of atoms.) Therefore it is necessary to consider the widely different methods of measurement. The uniformity of results leaves no doubt that atoms actually exist.

One method of measurement involved radioactivity. Individual alpha particles[6] are emitted at high velocity by a variety of radioactive substances. They can be counted and accumulated. The accumulation is then identifiable as helium gas, and by measuring the amount

[6] Alpha particles are the nuclei of helium atoms moving with great speed. Eventually they are slowed down, stop and attract electrons to form helium atoms.

accumulated and comparing it to the number of alpha particles counted, one can determine the actual size of helium atoms.

Another measurement was related to Faraday's discovery that certain amounts of electricity are associated with certain masses of matter. If there is a minimum size of matter, the atom, then there also should be a minimum electric charge, and all other charges would be simple multiples of this "atom" of electricity. An American physicist, Robert Millikan, found these minimum electric charges by investigating the behavior of tiny oil droplets. By observing the speed with which the droplets settled, he could get an accurate measurement of their size and weight. By then applying an electrical field, he could sustain the oil droplets motionless in midair. At this point, the charge on the oil drop gives rise to the force sufficient to cancel its weight. Thus the electric charges were measured. They turned out to be multiples of an elementary charge. Using Faraday's observations, the mass of the atoms followed.

Still another measurement involved the irradiation of atoms in a crystal with short wavelength radiation such as x-rays. This measurement is based on a phenomenon observed in connection with visible light. Different forms of radiation, such as x-rays, radio waves and white light are all similar in that they are electromagnetic waves. However, they differ in wavelength, and one of them, white (and visible) light, is a composite of colors which can be separated by a prism. Visible light varies from red to blue depending on the wavelength.

As light falls on a regularly spaced set of reflecting lines (a grating), it is reflected in well defined directions. These directions depend on the wavelength of the light and on the ratio of the wavelength to the spacing of the grating. If light from a point source is shone onto a phonograph record held at a grazing angle to the light beam, the colors of the rainbow will appear. The grooves of the record act as a grating, and red light is thrown back at a different angle from the yellow, green or blue. This is an easy experiment to perform.

Actually this phenomenon is called diffraction and differs markedly from reflection in which the angles of incidence and reflection are always the same. The grating in diffraction separates colors from "mixed" white light because the waves in visible light reinforce each other as they leave the grating at various angles according to the wavelength of light. There is a fixed relation between the deflection angles and the ratio of wavelength to the spacing of the grating lines.

Max von Laue, a German physicist, suspected that x-rays have a wavelength several thousand times shorter than visible light. He used crystals, which are similar to superfine gratings. The diffractions patterns for the x-rays yielded detailed information about crystal structure, such as the relative distance between atoms in the crystal. But the distances between atoms could not be known without finding out the wavelength of the x-rays. Fortunately, very small angle x-ray diffraction could be observed using man-made gratings. The advantage of small angles is that when the wavelength is sent through at a glancing angle, the wavelength behaves as if the distances in the grating were narrower. Comparing the low-angle diffraction from these gratings with the high-angle diffraction from crystals made it possible to determine the size of the atoms directly.

The fact, as I mentioned, that energy is fairly evenly divided between the various molecules of a system is the basis of another method of measuring atom size. Only a very few molecules will attain high energies; the structure of our atmosphere is evidence of this phenomenon. Although atmospheric structure is complicated in detail,[7] the general features are relatively simple. Three miles above the earth, the density of air molecules drops by about a factor two and continues to drop every three miles by a similar factor. (This is called exponential behavior.) At 15 miles, the density is only about 3 percent of its value at sea level. This is a very simple demonstration that only a few molecules will ever obtain the excess energy required to overcome the earth's gravitation to this extent.

One can reproduce this atmospheric distribution of molecules in a test tube, but, instead of molecules, one uses the smallest particles that the best microscope can see. These are called colloidal particles. Suspended in water, their distribution is similar to that of molecules in the atmosphere. Some corrections must be made for the buoyancy of colloidal particles, but when this is done, a good picture of the atmospheric distribution on a reduced scale is obtained. Because colloidal particles are about a hundred thousand times heavier than molecules, they simulate the atmospheric distribution in centimeters rather than in miles. For instance, the colloidal density will decrease to about one-half in a few centimeters as opposed to the three-mile distance required in the atmosphere.

This colloidal experiment (which is quite difficult to perform accurately) makes it possible to determine the approximate size of molecules. Indeed, statistical theory shows that colloidal particles and molecules have the same excess energy. The lighter a particle is, the

[7] One of the complications is that atmospheric temperature varies non-uniformly with altitude.

higher it rises. Height times weight is a constant. The weight of a colloidal particle can be measured. By comparing the heights of the columns, one can then get weight of the molecules by simple calculation.

It is interesting to consider why one can't see atoms in a straightforward manner. The reason is that visible light has such a long wavelength that it just washes over the individual atoms. A similar situation is that an ocean wave is not affected in its course by the pebbles on the ocean floor. It is not until the wave meets something comparable in size that there is an interaction, a reflection. However, the wavelength of an electron is small enough to "see" the atoms. (Indeed, as we shall discuss shortly, electrons possess a wavelength.) With an electron microscope, some molecules become visible.[8] (See Figure 3.5.) The electron microscope can measure the size of a known molecule directly and accurately, much as an optical microscope measures the size of a microbe. Seeing is believing, and since people can see molecules, it is no longer possible to deny their existence.

Most of these different methods for determining the size of atoms and molecules were originally found in the first decade of this century. Belief in atoms is based on these various measurements, all of which give (to reemphasize the crux of the matter) the same weight for a given atom. The more recent ability actually to see the aggregates of atoms forming large molecules and atoms themselves is the capstone of the proof.

I should not conclude this section without mentioning the actual approximate size of atoms. Their order of magnitude is approximately one-hundred millionth of a centimeter, or an Angstrom unit (called Å). To visualize this unit consider a little finger (which is about one centimeter broad at the tip). Divide by 10,000, and one enters the world of micro-organisms, just slightly bigger than the wavelength of light. Divide again by 10,000, and one arrives in the world of the atoms. This seems very small. It is not. A dog can smell a single molecule. Atoms are not that far removed from human senses.

Two Sciences in Collision

There were good reasons for the initial objections of Ostwald and Mach, based primarily on the violation of the laws which hold for

[8] Actually, Albert Crewe of the Enrico Fermi Institute at the University of Chicago has even managed to see individual uranium atoms. Those among the readers who may have seen films of this work will agree that atoms, which appear as points of light, may not look too exciting, but they do appear to be quite mobile.

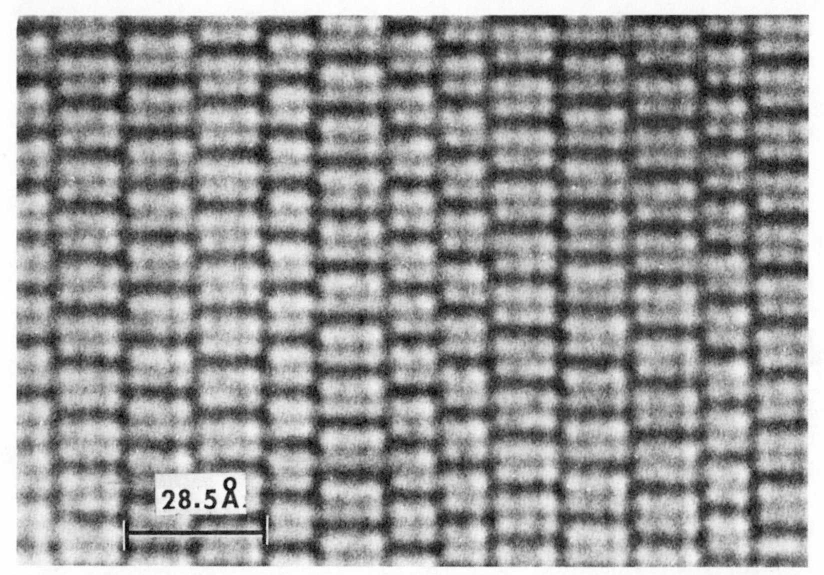

Figure 3.5. The Structure of a Niobium Oxide Crystal Seen Through An Electron Microscope.

This picture of a niobium oxide crystal was taken by Dr. John Cowley and Dr. Sumio Iijima at Arizona State University. The little black dots that seem to merge into lines are mostly oxygen atoms. The slightly heavier dots (rather difficult to discern) are niobium atoms. The white spaces are empty. The orderly nature of the crystal is clearly visible. The scale (28.5 Å or 28.5 x 10^{-8} centimeters) is indicated on the left. The diameter of an atom is little more than 2 Å.

macroscopic objects. Unlike the case in the previous chapter, where essentially the one experiment by Michelson and Morley disproved the concept of the ether and established the constancy of the velocity of light, the conflict in the case of atomic concepts was based on many issues. The very question of the existence of atoms produced a number of contradictions. Rather than a collision of two facts, there was a collision of two sciences![9]

In the first chapter, the reader encountered the science of classical mechanics (although in a somewhat indirect way). The science of the motion of bodies was fully accepted only when it could be applied to the motion of the earth and planets and to objects on the earth. Today no one doubts the validity of mechanics when it is applied to ordinary problems.

Chemistry—as applied not only to chemical reactions and syntheses but also to other neighboring branches of science such as

[9] A science in this sense is a systematic array of facts tied together by basically simple "laws," so that all the facts together have a power of conviction incomparably greater than any single fact could have.

crystallography, heat and parts of electromagnetism—had been so successful that no one could doubt its general concepts. But neither could anyone doubt classical mechanics. The existence of atoms led to a direct collision between the science of classical mechanics (and physics in general) and chemistry.

About the time that the size of the atom was definitely established, it was discovered that the atom was not truly indivisible. Instead, atoms were found to consist of a positively charged nucleus, very much smaller than the atom, with electrons loosely arranged throughout the volume of the atom. Because the negatively charged electrons were not pulled into the positively charged nucleus, it was assumed that the electrons moved around the nucleus like planets around the sun. However, moving charged particles should radiate energy, like an alternating current antenna. In fact according to classical ideas, one can calculate that—in less than one-millionth of a second—the electrons should radiate away their energy and "fall" into the nucleus.

Still more serious difficulties were encountered as more was learned about atoms and their behavior. Chemists had found that all atoms and molecules of a given type had identical properties. Yet if atoms and molecules were divisible, and they were undergoing tremendous numbers of high velocity impacts, why didn't they change their intrinsic structure and hence their properties?

The final blow came when it was realized that the classical theory of heat did not apply inside atoms and molecules. According to the statistical theory of heat, all parts of the atoms and molecules should accept some heat energy, but the electrons within the atoms and many atoms within molecules seemed not to do so. (This limitation, as shall be seen later, is not the only one in the statistical theory of heat.) The fact that some parts of atoms and molecules do not respond to heat energy was the starting point of quantum mechanics. It was closely connected with the first quantitative success in clarifying the atomic paradox.

One remarkable example of this difficulty is mentioned by Willard Gibbs. Gibbs was worried about an exception to the predicted behavior of diatomic molecules. For instance, atomic nitrogen has three directions in which to move: forward and backward, up and down, and left and right. Therefore, diatomic nitrogen ($\mathbf{N}_2$) should be capable of six different types of motion, the three mentioned above (for the molecule as a whole) plus the two rotations shown in Figure 3.6., and finally the oscillation of the two nitrogen atoms alternately increasing and decreasing their distance. However, it

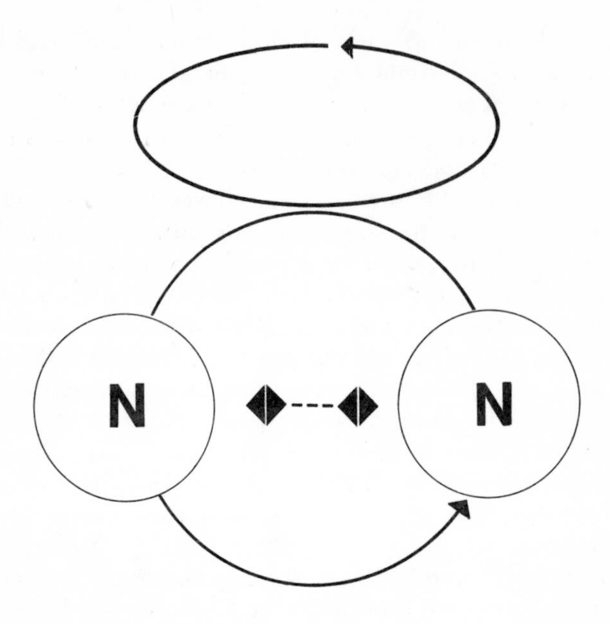

Figure 3.6. The Rotations of Diatomic Nitrogen.

The oscillation along the nitrogen-nitrogen axis, which occurs only at very high temperatures, was not observed during Gibbs' time. Two independent rotations of the molecule around two perpendicular axes, which are also perpendicular to the line connecting the two **N** atoms (shown in the figure), do occur at normal temperature.

appeared from measurements that only five of the six possible motions share the available energy. In particular, the oscillation does not occur. This absence was a complete surprise in terms of classical physics.

This inconsistency deeply disturbed Gibbs, as he made clear in the introduction to his famous book on statistical mechanics.

> It is well known that while theory would assign to the gas six degrees of freedom per molecule, in our experiment on specific heat we cannot account for more than five. Certainly, one is building on an insecure foundation, who rests his work on hypotheses concerning the constitution of matter.
>
> Difficulties of this kind have deterred the author from attempting to explain the mysteries of nature and have forced him to be contented with the more modest aim of deducing some of the more obvious propositions relating to the statistical branches of mechanics. *Here, there can be no mistakes in regard to the*

agreement of the hypotheses with the facts of nature, for nothing is assumed in that respect. The only error into which one can fall, is the want of agreement between the premises and the conclusions, and this, with care, one may hope, in the main, to avoid.[10]

Gibbs appears to be thoroughly worried. He heaps qualification upon qualification hoping that he (a truly great mathematician) did not err in this work which he based firmly on the foundation of mathematics. That he describes the real world, he hardly hopes. It is a peculiar situation when the author of a rigorous treatment of statistical mechanics does not dare to consider his results as representing the physical world but wants to call them part of pure mechanics, treated as a branch of mathematics.

Quantum Theory: An Attempted Compromise

The lines quoted were written in 1901. Unknown to Gibbs, the first relevant step had just been taken to solve this and all similar problems by a German physicist, Max Planck. Planck was interested in the application of the statistical method to the problems of radiation rather than to the problems of particles. Planck's work was pertinent because not only the particles of a system share the energy of heat but also all the various forms of radiation share this energy. The simple classical rule is that all possible wavelengths share equally the energy that is available.

Planck noticed that this rule is satisfied as long as the wavelength of the radiation is sufficiently large. However if one tries to apply this rule to all wavelengths, an obvious and serious difficulty occurs: there is no clear limit of how short an electromagnetic wave can be. Ultraviolet light is shorter than visible light; x-rays are shorter than ultraviolet; gamma rays are shorter than x-rays, and so it goes without limit. If all wavelengths were to share in the energy, then they could take up an infinite amount of energy. The classical statistical approach led to a clear absurdity.

Planck found an empirical solution to the contradiction by trial and error. At first his proposal seemed to make sense to no one, not even to Planck himself. He simply stated that energy cannot be obtained in arbitrarily small units. Planck's *quantum* view of energy was that it behaved rather like money. In the world of atoms, being wealthy corresponds to having lots of energy. Rich people are seldom

[10] Emphasis added.

bothered by the limitation that one cannot spend or receive money in units smaller than a penny, but the limitation does exist. It is most surprising that a smallest unit—a sort of penny—exists in energy transfer as well as in money.

The situation becomes even more unusual when one looks at the details of Planck's proposal. The smallest unit of energy is not always the same. Planck defined a unit of energy in terms of periodic motion.[11] He found that the minimum unit of energy that can be transferred is proportional to the frequency of the periodic motion. The larger the frequency,[12] the larger the minimum unit of energy. The equation that describes this situation is

$$\mathbf{E} = \mathbf{h}\nu,$$

where $\mathbf{E}$ = energy, ν = frequency, and $\mathbf{h} = 6.625 \cdot 10^{-27}$ erg-seconds, Planck's constant of proportionality.

Planck's assumption explains at one stroke all the peculiarities that have been discussed. Frequencies in radio waves are quite low, so the quantum ($\mathbf{h}\nu$) is quite small compared to the available energy. Therefore, Planck's restriction plays no role. As the frequency moves from this low level to the frequency of visual light, the quantum of energy becomes bigger (though light is basically not different from radio waves). The "penny" of radiation necessary for emission of visible light is too big for our daily transactions since wealth is determined by temperature. Normal temperatures do not cause a body to glow. When the temperature is raised to "red hot," red light (which has the lowest frequency among the visible colors) will be emitted. At white heat (which is, of course, higher), all colors are produced and give the standard mixture of white light.

Similar statements hold true for vibrations in solids where many vibrations have such a high frequency that the smallest acceptable unit of energy doesn't come into play at normal temperatures. In the case of the nitrogen molecules, the missing mode of motion is due to the fact that the smallest quantum of energy which nitrogen could accept is hardly ever available unless the gas is extremely hot.

Finally, inside the atom, electrons are moving around the nucleus at very high frequencies, in general higher than that of visible light. The energies available at ordinary temperature have no effect, and the atom acts as though it were rigid, as though it had no moveable

[11] Periodic motion is movement which is repeated with regularity. Examples include the motion of a pendulum or the movement of planets around the sun. Wave motion also is periodic as the wave form repeats. More precisely, the periodic motion in Planck's definition is "sinusoidal;" that is, the same motion that a pendulum of small amplitude performs. (See Figure 3.7.) What mattered here for Planck was that this is the type of variation with time that occurs in light of a sharply defined color.

[12] Frequency and wavelength, though connected, are, of course, quite different. (See Figure 3.7.)

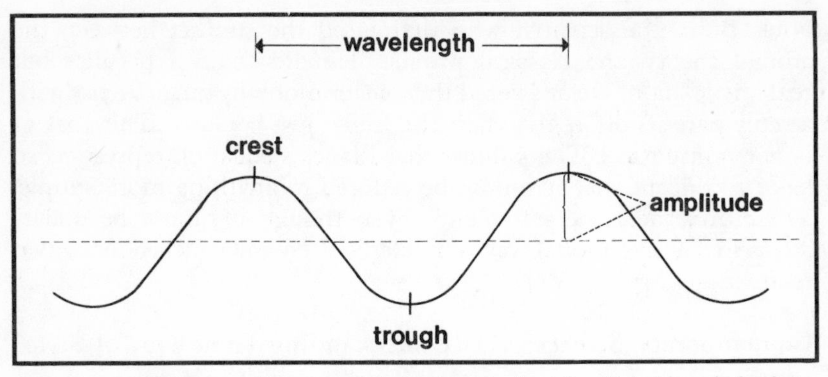

Figure 3.7. The Sinusoidal Wave and Wave Descriptions. In old-fashioned physics, a wave is a disturbance which moves through a medium, transmitting energy. If one were to tie a cord to a doorknob, pull the cord taut and wiggle the end of the cord, one could see some of the characteristics described in this illustration.

A wave has a length which is measured from crest to crest, and an amplitude which is measured from a baseline (halfway between the crest and trough). The wavelength of red light is about twice that of violet.

The frequency of a wave is the number of crests that would pass a given point in unit time. The wave number is the number of crests that occur in unit length.

Wavelength times frequency is the common wave velocity. In general, as wavelength increases, frequency decreases; as wave number increases, frequency increases. Amplitude in classical physics is connected to the amount of energy the wave carries. (Energy is proportional to the square of the amplitude, unless the amplitude becomes very large.) Amplitude of wavefunctions in quantum mechanics has a very different meaning.

parts within it.[13] Planck's hypothesis was the first step toward understanding why atoms don't collapse, why there isn't infinite energy present in radiation and why atoms and molecules don't change their intrinsic structure on impact.

There is a parallel that may be drawn between these ideas and the ones presented in Chapter Two. If one starts with a simple idea, such as the invariant $(\mathbf{ct})^2 - \mathbf{r}^2$ or the equation $\mathbf{E} = \mathbf{h}\nu$, a great number of peculiar consequences follow. Planck's equation is even simpler than the invariant, but at the same time it is even more surprising because it connects two quantities that originally (and in terms of common understanding) had nothing to do with each other—energy and frequency.

Why do various atoms accept only specific quantities of energy? Planck gave the correct answer in the form of an equation but stopped short of discussing its meaning. The answer came from

[13] A rigid object (which is actually a somewhat loose description) transmits rather than absorbs energy. The difference in absorption of energy that exists between rigid and non-rigid objects is seen in the differing results of hitting a billiard ball with another billiard ball or hitting it with a bean bag.

Niels Bohr, the scientist who first faced the conflict between the atomic theory and classical physics. He did so in a peculiar but realistic fashion. He answered the question of why quanta exist with a reply parents often give their children: "just because." This answer is not nonsensical. The point is that Planck's equation represents so basic a concept that it cannot be reduced to anything more simple. Therefore, the basic statement—even though it cannot be understood in a conventional way—should be used and conclusions drawn from it.

Contemporary Science: Has Understanding Gone Out of Style?

Does this mean that we should despair of ever understanding the things we discuss? I remember an incident that took place about a third of a century after Planck's remarkable proposal. The consequences of the proposal had been crystallized, and a number of philosophers were discussing the quantum theory (which by that time was fully developed) with Niels Bohr at a meeting in Copenhagen. The exchanges were all very pleasant and friendly, but after the philosophers had finished agreeing with Bohr and gone home, Bohr appeared deeply depressed. Someone asked him why he was dissatisfied with the discussion. I remember his response clearly. "If someone talks about Planck's constant and does not feel at least a little dizzy, he has not understood what he is talking about!"

Still, understanding is something very personal and somewhat peculiar. I mentioned my operational definition, and there is a worthwhile point to it. To understand atomic theory (or relativity), one needs to draw as many conclusions as possible; one needs to find that there are no contradictions between those conclusions and that all relevant experiments are explained. If one becomes so well practiced that the consequences can be guessed before they are calculated, then one can and should say that the new theory is understood.

Despite the generally accepted statement that it is neither necessary nor possible to visualize the atomic process in conventional terms, it is useful to try to do so. At least this exercise may extend one's ability to correlate what is new and strange with common experiences.

One of the difficulties with Planck's equation is that it applies only to periodic motion. Bohr recognized that, in the case of particles, there is a connection between the motion of the electrons in an atom and the frequency of radiation that the atom can emit. It is dangerous to try to form too definite a picture of electrons moving about the nucleus, but Bohr introduced the concept of stable "stationary" states within atoms (a concept directly taken over from chemistry where atoms are conceived as being in definite states). Bohr used this concept to formulate the idea of light being emitted when electrons "jumped" from one state to another.

Each state has a definite energy, and the difference between the energies of two states, called **E**, was used to find the frequency, ν, of the light emitted or absorbed in the transition by use of Planck's relation, $E = h\nu$. Generally the stationary states are widely spaced but get closer together as forces and frequencies get smaller (in highly excited states). In this region, the rules of quantum theory gradually merged into or "corresponded" to the rules of older physics. Bohr used this "correspondence principle" to guess the rules of quantum theory.

Dualism: Waves and Particles

The question could not be avoided: Is light a wave motion? More particularly, if light contains quanta of energy, does light consist of particles? Actually this question is quite old. It goes back to an historic debate between Isaac Newton, who believed in the particle nature of light, and Christian Huyghens, who believed in the wave nature. This debate had a most peculiar semiscientific flavor. Preconceived ideas played a most important role in this discussion because the strongest arguments on both sides were as yet unavailable.

It is particularly interesting to trace back this dualistic interpretation of light into the old elementary facts concerning the behavior of light. Newton observed that light is sharply bent or "refracted" when it enters water from air. He ascribed this behavior to the attraction of the light particles from air into water. Huyghens explained the same phenomenon by the use of wave theory. It turns out that the two explanations correspond to each other as if they were the same story told in two languages. One only has to find the correct translation: momentum (mass times velocity) in the particle theory corresponds to wave number (the number of waves per unit length) in the wave theory. The two explanations become identical if one postulates that momentum and wave number are proportional to each other. This is a statement quite similar to $E = h\nu$, where energy and vibrations per second turn out to be proportional. (See Appendix VIII.)

A second elementary observation bearing on wave-particle dualism is the Doppler effect. As a simple example, consider light reflected perpendicularly from an approaching massive mirror. According to the particle theory, the reflected particles have a higher energy, just like a tennis ball hit by an approaching tennis racket. According to the wave theory, the image of the light source approaches with twice the velocity of the mirror, and an approaching source of waves gives a higher frequency to the waves. It is easy to understand this: the result of the approach is that wave crests will follow each other in shorter time intervals. For example, the higher frequency in sound (this is represented by higher pitch) is easily heard if one listens to the horn of an approaching train.[14]

The strange fact is that, again, the effect of the approaching mirror can be explained by the particle theory or by the wave theory of light. The two give compatible results when one assumes that the energy of the particles is proportional to the frequency of the waves, and that momentum and wave number are proportional. Furthermore, the constants of proportionality must be the same. (See Appendix IX.)

Thus we find a most remarkable situation. Planck had to make assumptions of finite energies or quanta for light, thus raising the suggestion of a dual particle-wave nature of light. Now it can be seen that this dual nature is indeed compatible with simple facts concerning light only if a proportionality such as $\mathbf{E} = \mathbf{h}\nu$ holds. What I am saying seems strange, but its mathematical form is simple.

Planck offered the first evidence for the particle theory of light two centuries after Newton. This was soon followed by Einstein's observation that light, well established as a wave motion, nevertheless delivers energy to electrons in packets or quanta of the size $E = h\nu$. In 1905, Einstein looked at light waves as particles; in 1924, de Broglie considered the particulate electrons as waves.

The arguments for the wave hypothesis of light are much older than those for light as a particle. First came Thomas Young's "interference" experiments which demonstrated in 1801 that light plus light can give darkness at certain, well-defined locations. Waves can perform this trick, but particles cannot. Then Maxwell's discovery of magnetic and electric fields seemed to complete the proof—showing that light consisted not of just any kind of waves but specifically of electromagnetic waves, that is, of oscillations of electric and magnetic fields.

Beginning with Planck, we have the unprecedented situation that

[14] A popular story among physicists involves the excuse used by a driver who ran a red light. He explained the Doppler effect to the arresting officer and claimed that due to this effect, he had seen the light as green. The officer nodded, tore up the ticket and wrote out a new one—for going 300 million miles per hour over the speed limit.

light has *both* a particle nature and a wave nature. It was Bohr who finally took the position that light has a dual nature just as human beings do. Description of light as a particle or a wave is necessarily incomplete, just as man cannot be described completely either as an assembly of matter or as a pure spirit. The difference is that in this latter age-old paradox, we still lack the answer as to what man is, while in the case of light and the atomic world, a mathematical and consistent description has been provided. This description does violate our intuition, but nowhere does it violate logic. (Most people find it hard to keep their most elementary intuitions separate from logic.)

If light has a dual nature, why should the same statement not hold for the constituents of atoms, the electrons and nuclei? Indeed, why should atoms, molecules or aggregates of molecules be different? They too may have a wave nature in addition to their old and more plausible particle nature. This daring hypothesis was made by a French duke, Louis de Broglie, when he was a young man. A remarkable story attached to his discovery tells that when de Broglie handed in his thesis proposing this at the Sorbonne, the professors were unanimously astounded. The thesis appeared to make no sense at all. It asserted, as I have mentioned, that electrons have a wave nature. Because electrons are constituents of atoms and atoms are particles by definition, it seemed nonsense to assume that electrons were waves.

There's little doubt that de Broglie's idea would have been rejected with ridicule except for a few difficulties. De Broglie was a duke, and his father had been a powerful prime minister of France. Under these conditions, rejecting a thesis required a truly unshakable conviction. The professors were fortunate in that a notable physicist was visiting. With relief, they offered de Broglie's thesis to Albert Einstein for comments.

Einstein read the thesis and was enthusiastic. He saw in this idea a thought analogous to a proposal he had made regarding light some nineteen years earlier. He immediately spread de Broglie's concept to the rest of the physics community (as well as recommending that the thesis be accepted). One member of the larger physics community, Erwin Schroedinger, not only grasped the importance of the idea but used it to produce a formulation that is applicable to a wide variety of problems.

Schroedinger's view was that if electrons have a wavelength, then these wavelengths should fit properly into the atom. If we consider a circle with waves "moving along" the circle, the waves should "close" or "wind up" on themselves in one circuit. Otherwise, one has

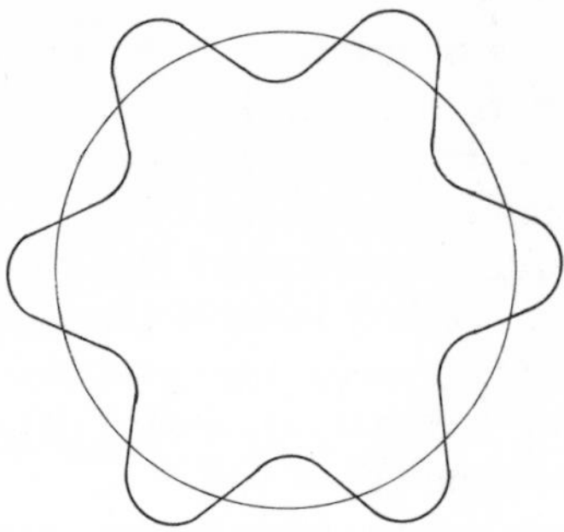

Figure 3.8. A Visualization of de Broglie's Wound-Up Wave. The light circular line corresponds to the "boundary" of the atom. The heavier curve represents the wave path of the electron circling the nucleus of the atom.

to continue several times around the circle with several amplitudes corresponding to each point on the circle. In this case, the amplitude at each point would total zero, and there would be no wave. (See Figure 3.8.) As mentioned earlier, Bohr had postulated, in agreement with the findings of chemistry, that atoms exist in definite states. One can see here that this corresponds to the idea that the possible wavelengths of the electrons must fit into the dimensions of the atom. All this can be described mathematically and accurately with the aid of differential equations from which one can determine a *wave function* (represented by Ψ) that describes the behavior of the electron.

Schroedinger's work on the atoms produced the most remarkable results. He was able to write differential equations that explained the quantum behavior of matter. Indeed, they explained in principle all of chemistry. Unfortunately these equations are complicated enough that scientists can't readily solve them. In the simpler cases where solutions are available, the agreement with measurements is excellent. In more complicated cases, extremely useful approximations have been developed. These allow a much deeper understanding of general laws that chemists have known for a long time.

It turned out that the collision between chemistry and physics resulted in a unification of the two into a single and most impressive science which appears to explain almost everything that can be observed about ordinary matter. It does so, however, at considerable expense to customary, somewhat naive visualizations.

Looking at the wave interpretation of electrons which Schroedinger provided, one is confronted with the sort of dilemma that confronted the physicists of the previous century in considering light waves. Waves usually bring to mind the vibration of a medium. However, the wave nature of the electron moves in no such medium, and the visualization becomes absolutely impossible when many electrons and additional particles are to be considered. In this case, the wave motion is mathematically described in multi-dimensional space.[15] The statement may seem unusually complicated at first, but ultimately it appears both simple and logical.

If the electron is going to be a wave, one has to assign it a function, that is, an amplitude of the wave. This function will depend on the position of the electron (the familiar **x, y, z**) at each moment. This amplitude is obtained as the solution of a differential equation, appropriately called Schroedinger's equation. At this point the description of electron waves do not essentially differ from the description of sound waves. However, the amplitude, as I mentioned before, does not represent the displacement of a medium. Rather the amplitude is related to the probability of finding the electron at any given position.

According to Schroedinger's picture, the electron is not a point particle, but is "smeared out" to the extent that it is easier to think of it as a "cloud" surrounding the nucleus of an atom. This cloud has a density, and the density is defined as the square of the amplitude.[16] Now it is possible to describe the probability of finding the electron at point **x, y, z.** The formula is

$$|\Psi|^2 \, dxdydz$$

where Ψ is defined as the wave function, and **dxdydz** is a tiny volume about the points **x, y,** and **z.** (See Appendix X.)

Physics Has Gone Off the Gold Standard

I have discussed probabilities previously, but the probabilities now being considered are of a fundamental nature. They are not being used in the sense of the probabilities associated with life insurance or with the statistical nature of heat. These probabilities are now expressing fundamental uncertainties that cannot be removed—even in principle. That means that neither more detailed calculations, nor more careful measurements, will ever give a more certain knowledge. This is something fundamentally different from the probabilities used in the classical world of statistics.

There is an analogy which describes the quantum situation so ingeniously and attractively that I want to use it even though it is

[15] A single iron atom requires 81 dimensions because the function depends on the **x, y, z** coordinates of each one of its electrons and also on the three coordinates of its nucleus. This means 81 coordinates, or for a mathematician, 81 dimensions.

[16] More exactly, the density is defined (for the mathematician) as the absolute square of the amplitude. The wave function is complex, and the absolute square is real and positive (or zero).

somewhat lacking in accuracy. It was used by an English astronomer and author, Sir Arthur Eddington. In a lecture he gave in 1932, Eddington compared the old and new statistics with a contemporary situation in economics.

A rich man with too much gold to carry around can deposit it in the bank. He can easily operate with bank notes because everyone is convinced that there is gold backing up the notes. In physics, the ability to calculate the exact motion of atoms in a gas using classical mechanics corresponds to the gold. Everyone is willing to accept the statistical calculations under the classical theory because it is theoretically possible to make a more detailed explicit calculation. So long as banks were on the gold standard and physics was using classical mechanics for the statistical basis of calculation, everyone could have complete confidence in the state of affairs.

What happened is that with quantum mechanics, physics went off the gold standard. (In 1932, the Bank of England went off the gold standard as well!) One still had the statistics (the bank notes), but the detailed calculations (the gold) behind the statistics were gone. The calculations now represented the reality of the situation in and of themselves. They were no longer just a shortcut or a simpler way to do things.

Eddington's example explains that no special meaning can be attached to the equations that enable one to calculate the probability amplitudes. There is meaning in the following two facts, however. First, the calculations of probabilities give excellent agreement with measurements in a very wide range of physical situations. Second, these calculations can be used to predict phenomena not previously observed, both qualitatively and quantitatively. These successes are surely not trivial.

It was hard for scientists to accept this situation as a final answer. One man very reluctant to believe it was a person who contributed an immense amount to the initial development of wave-particle dualism: Albert Einstein. When he saw the "finished product," when he was confronted with the intrinsic nature of the uncertainties, he balked. He stated his objection succinctly: "I can believe that God governs the world according to any set of laws, but I cannot believe that God plays dice."

A second more significant argument is that if one introduces probability without recourse to checking in detail, one loses the assurance that what is seen, sensed, or discussed has anything to do with reality. After all, it is through cause and effect that the external world produces impressions on our minds. If one denies causality, one may have denied observations.

I do not find this argument totally convincing. Statistical laws are also very definitely laws. Furthermore, they are laws which can be checked, not in the individual case, but sufficiently to establish at least a type of loose causality. In fact, this situation may be more in agreement with what we know by introspection about ourselves than causality.

Measurement and Reality

Bohr and Heisenberg elucidated the meaning of statistics as it applied to atomic theory. According to them, the discipline of quantum mechanics is put together from two different phases. The phase that describes the evolution of a wave function, signifying probabilities, was Schroedinger's work. The other phase is the measurement of the properties of an atomic system, for instance, the position of an electron.

The measurement process which determines the configuration of a group of electrons proceeds independently of strict classical causality. In general, there is no certainty as to which of the possible configurations will be found when one actually measures. Only a probability can be calculated. This limitation is not just ignorance but is an inherent feature of our world.

The measuring process itself is not governed by the cause-effect relationship, or as mathematicians would say, "it is not to be described by differential equations." Indeed, an important role of differential equations in physics is to start from a detailed description of a situation at one instant in time and then calculate, using the rules of causation, the situation at the next instant where the time difference between the two instances is very small. All this is replaced in the measuring process by a choice, performed by the measurement. Repeated measurements on similar systems can only check that the probabilities were correctly calculated. The measurement can't be analyzed in time. Neither can one answer, for instance, the question of how an electron is found at a definite spot when, prior to measurement, the probability of finding the electron was distributed over an extended region.

A more intricate portion of the Bohr-Heisenberg discussions dealt with the problem of why it can't be decided, by some experiment, whether an electron (or light) should be described as a particle or a wave. At this point, the discussion went further and answered the question of whether some other concept than particle or wave could clarify the problem. The answer was: NO!

Bohr never tired of emphasizing the fact that the foundation of science is a straightforward direct ("classical") observation. Otherwise, there would be no way of saying what one was discussing. It is not possible to talk about measurement unless it is described in classical terms.

Yet now, these measurements will lead to uncertain results. There is no real contradiction in this point. The contradiction is avoided by noticing that one is trying to make very fine measurements where the measuring instrument interferes with the quantity that one is trying to measure. Bohr was fond of explaining this by analogy. The question *Are you asleep?* can be answered only at the price of waking up. The probing of whether one is asleep interferes with the state under observation.

But, accepting this point of view, what effect does this dilemma have on experiments designed to verify whether an electron is a wave or particle? In particular, one may perform an interference experiment, using an electron beam. To obtain interference, electrons must be represented by waves. (See Figure 3.9.) The lines in the figure represent wave crests, while the shaded areas show two screens, the one on the left having two holes in it. The theory predicts that as the waves arrive at the second screen (on the right), the wave action will cancel at the points where crests and troughs coincide (which is, indeed, a demonstration of wave *interference*).

As yet, no one can execute this experiment with electrons, though an essentially equivalent experiment in which electrons are scattered from the regular lattice of atoms in a crystal has been carried out. The figure shows a "thought" experiment (Gedanken Experiment) in which the *idea* of interference as related to electrons is emphasized.

Since the experiment on the crystal has been carried out, there can be no doubt that the simplification of the experiment depicted in the figure is not essential. One can therefore assert that electrons must be represented by waves rather than by particles because it would be absurd to say that at certain spots on the second screen, no electrons (considered as particles) ever arrive merely because they had a choice of getting there through one hole or the other.

But the disproof of the particle theory is not complete unless one can trace from case to case through which of the two holes the electron-particles found their way to the second screen. The essential point in the Bohr-Heisenberg argument is that whenever one tries to establish which of the two paths the electron took, one disturbs the electron to a sufficient extent that the interference pattern is destroyed. Whenever the electron is put to the crucial question,

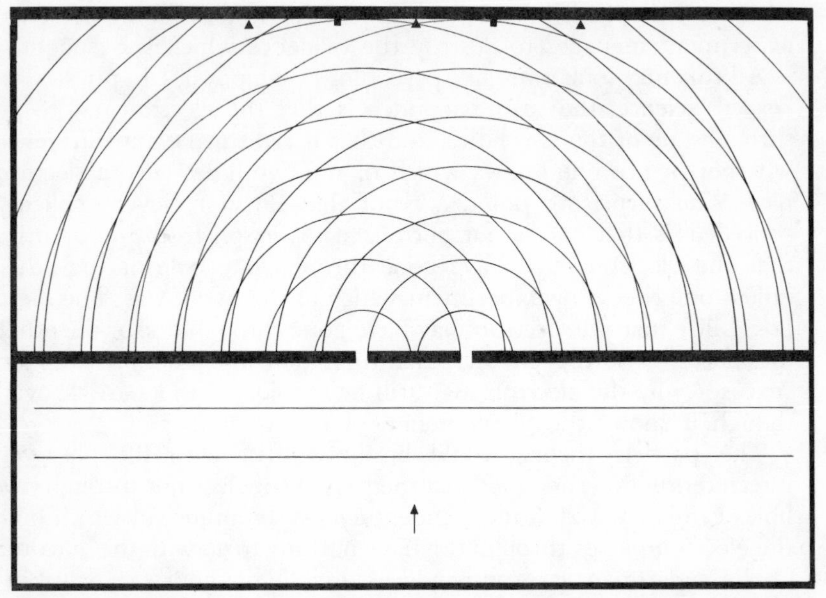

Figure 3.9. Interference Due to Wave Motion. This figure describes wave interference, an experiment which can be carried out directly only with light waves, but which can be applied to electrons as a thought experiment. The wave crests (of light or of electrons from an electron beam) are represented by the lines, and the troughs lie at the midpoint between two lines. Waves originate at the bottom and fall upon a plate in which there are two holes. The waves spread out on the far side of the plate along circles. The intensity of wave motion varies as it reaches the second screen shown at the top.

The dots shown on this uppermost screen represent the places of maximum intensity (positive interference) and the triangles, the points of lesser intensity (negative interference). At the point directly opposite the center of the plate between the two holes, two crests occur simultaneously, and the positive interference is at a maximum. At the next dot (down or up), adjacent crests coincide; at that location a very short time previously, two troughs have coincided. Correspondingly, secondary intensity is maximal here (again because of positive interference).

At a location approximately midway between these two dots, but slightly closer to the central point, the distance from the nearest crests adds up to precisely one-half wavelength. (This location is marked by a triangle.) A short time before the pattern shown in the figure occurred, one would have found a crest and a trough coinciding at this point. The two cancel, and one can show that they continue to cancel at all times (negative interference). The dots and crosses alternate, corresponding to positive and negative interferences. These are called the interference fringes.

In an actual three-dimensional situation with a light interference experiment, the central dot would be seen as a bright region surrounded by alternating circles of light and darkness. There is no way to explain the negative interference (the dark areas) using the particle theory. One would be forced to say that when electrons or light quanta have two paths to get to the same point, practically nothing arrives—which, of course, is ridiculous.

whether it is particle or wave, the process of questioning by actual experiment manages to destroy the evidence which was sought.

All this may look more like psychology than physics. In physics, an "exact" science, the customary idea is that the electron had to go through one of the two holes, and that it can't make any difference whether or not one knows which of the two holes let the electron pass. The surprising point, a point that Einstein never explicitly conceded, is that one is not permitted to reason from the premise that the electron must have gone through one hole or the other unless one checks by experiment which of the two it was. The mere possibility that the electron may have gone through one or the other influences, absurdly enough, the final outcome and provides an "excuse" why the electron may still be considered as a particle even though it shows the phenomenon of interference.

The particle theory would be in conflict with the observed interference if one assumes that the passage of electrons through two holes can be viewed as independent events. Ignorance of which hole the electron passes through can have nothing to do with the outcome of the experiment, as one would assume. This ignorance or uncertainty is, however, crucial. It should be recognized that this situation is in sharper conflict with what is usually considered reality and common sense than the theory of relativity. Einstein's innovation merely redefined the concepts of time and space—and redefined them in unambiguous terms.

It is the ambiguity in the dualistic explanation that continued to disturb Einstein and many others. According to their older view, full reality is inconsistent with ambiguity. This, however, is too narrow a view of reality. The physicist may invoke tradition when he protests that in actuality the electron must have gone through one hole or the other, and one's knowledge of which choice has been made can have nothing to do with the outcome. Yet, we are willing to accept in observing ourselves (and by implication, others) that such paradoxical situations as love-hate relationships exist.

Even if it is conceded that a love-hate relationship can be resolved by reasoning or other means, it should also be immediately conceded that once this relationship is resolved and reduced to love or hate, the outcome in one's actions will be quite different from those that would have resulted from the original love-hate relationship. Is psychological ambivalence real? Is the path of an electron less so?

Whatever way one tries to find out the position of the electron (or another of its properties), the measuring apparatus is itself subject to laws of physics—thereby affecting the reality one seeks to observe—

and can be included in the description of the experiment. One may consider this problem from two extremes. The first is to describe the electron as subject to the peculiar laws of dualism and consider the measuring apparatus (and everything else) as described by "classical" laws.

The other extreme is to attempt a description of the electron, the measuring apparatus, and all the rest of the universe as subject to the new laws of dualism, restricting the classical description to one's own final recognition of the outcome of the experiment. Bohr and Heisenberg argued that the line must be drawn somewhere because the final step of one's own recognition cannot be based on anything but "classical" common sense.

A further point is that, within very wide limits, it makes no difference where one draws the line of division between dualistic description of atomic theory and classical description of the observer. It is simpler, however, to include as much as possible in the classical description and as little as possible in the portion described by characteristic atomic laws.

Uncertainty Becomes Science

This intrinsic uncertainty is made concrete by the Heisenberg principle. Basically it says that one cannot simultaneously know the position and the velocity (actually the momentum) of an atomic particle. Knowing the position exactly means that the state function is infinitely narrow about that point. However, the state function then predicts *any* value for the velocity of that particle. Similarly, exact knowledge of the velocity results in a state function that predicts that the electron may be found *anywhere* with equal likelihood. (See Appendix XI.)

In the interference experiment described above, a determination as to whether the electron has passed through one hole or the other must give a more accurate description of the position of the electron. This can be gained only at the price of a loss of knowledge concerning its momentum and, therefore, concerning the direction in which it will continue to move. Therefore, one can no longer predict according to the particle theory where it will land on the second screen.

Heisenberg pointed out that causality is not so much violated by the uncertainty principle as causality simply cannot be applied here. Causality applies in classical physics only if both the position and the velocity of the particle are known in the same instant. Because this knowledge is limited, the validity of causality is correspondingly limited.

Once quantum concepts were firmly established, the temptation arose to apply these ideas to everything, to macroscopic objects as well as to atoms, to measuring apparatus and the object measured, even to people. Everything in the world has wave properties and particle properties according to this view. However, due to the amount of mass humans have, the uncertainty principle imposes quite negligible limitations when one's own position and velocity are to be determined.

There is something most consistent in uncertainty. The consistency is that the uncertainty relation applies to all objects in the universe. This must be true because if there were an object whose state could be determined more accurately than is allowed by the uncertainty relations, then that object could be used as a measuring instrument to determine the relevant quantities of other objects. Just as the existence of ether would have solved the problem of who is at rest by providing a measurement standard, so an object which has known values of position and momentum would provide a standard for other measurements. If the uncertainty principle does not apply to everything, it applies to nothing.

Having considered this sober limitation, I want to mention an amusing historical fact. Heisenberg, to develop the ideas for the measuring process, used concepts related to the ordinary optical microscope. He certainly had to be fully familiar with the power of resolution of a microscope to do this. However, he couldn't answer a question on this subject during his Ph.D. examination. Louis de Broglie received a Nobel Prize for a thesis that was nearly rejected, and Heisenberg got his Nobel Prize in part for work that almost penalized him in his thesis examination.

Einstein and Bohr had a number of important discussions regarding the measuring process, uncertainty relations and statistical interpretation. Einstein was ingenious in trying to disprove the uncertainty relations, but they stood up to his tests.

There are still attempts to return to a non-statistical fully causal theory by instituting the "hidden variable" concept. This concept involves the reduction of atomic theory to a microstructure beyond experiments. It is hard to prove a negative, so a completely convincing proof—that this reduction can never be accomplished—has not been provided. The reduction has been attempted a number of times, but the result has always been a great amount of added complexity with no additional insight. The principle of the pursuit of simplicity effectively rules out such an approach unless there are experimental findings that force one to go beyond current ideas. Chances are such findings would demand moving forward, not back to classical or even semi-classical concepts.

Complementarity: Creating a Double Anchor

Bohr attached a great significance to quantum mechanics. From his very earliest work, he seemed to have realized what was coming. He wrote repeatedly about new ideas loosely connected with atomic theory. The most important of these is called complementarity. This word suggests that no single viewpoint is necessarily more valid than another opposite one. Complementarity states that one cannot get an objective and complete understanding in many situations unless one starts from two (or more) approaches that appear to be mutually exclusive. In some ways, this idea is a very old one, but it is now applied to some of the most basic ideas, like the position of a particle.

Complementarity is closely connected to the idea of dualism. In the 19th century, there was a sharp distinction between the view of humans as bodies and the view that humans are souls. Bohr's complementarity would have been considered obscurantism. At that time, one was not allowed to say that we are both. Bohr asserts, correctly I believe, that unless we consider ourselves from both points of view, we have an incomplete idea of what we are. In the world of the atoms, at any rate, complementarity is compatible with a science that is free of contradictions.

The question of whether light and electrons are waves or particles cannot be decided. What is true is that when light is observed as waves, it cannot simultaneously be observed as particles, and vice versa. Bohr's concept greatly enriches our ability to look at the world, and it will undoubtedly continue to be useful in the future. At the same time, it sacrifices the unrestricted applicability of causality. Perhaps this is something which is worth sacrificing.

One of the most peculiar and satisfying aspects of atomic theory is that in some ways it brings the methods of such widely differing fields as those considering the behavior of an atom and the behavior of a person closer together. But this statement must not be misunderstood. I do not mean to claim that mankind has arrived at an understanding of life or consciousness. All that can be claimed (and Niels Bohr did claim this) is that when and if life or consciousness is understood, the methods that were used to understand the atom will be, in some new manner and on a different plane, needed again.[17]

The relationship of Bohr to atomic physics was quite a peculiar one. His name brings to the minds of people the atomic model, a model completely discredited by Heisenberg (who was his student) and by Bohr himself. But one must remember that when Bohr proposed this model, he emphasized that it should not be understood in a concrete way but rather in a symbolic manner.

There is a particularly characteristic story about Bohr's relationship to atomic physics. The incident supposedly occurred about 1928 when, due to the work of de Broglie, Schroedinger, Heisenberg, Bohr himself and others, the theory of quantum mechanics was completed. One of Bohr's friends asked him: "What are these new ideas about the nature of atoms?" Bohr explained (placing greatest emphasis on the paradoxical points). When he finished his lengthy discussion, his friend said with amazement: "But this is precisely what you told me in 1912." The novelty was obscured because the mathematical formulations were inaccessible to Bohr's non-scientist friend and also by Bohr's way of talking. However, the story accurately points out that Bohr had, indeed, anticipated on an intuitive basis much of what the resolution of this theory demonstrated.

It is also relevant to record that Bohr emphasized, even demanded that the idea of complementarity—that it is necessary to consider the same situation from two mutually exclusive points of view—should be taught to all eighteen year olds. He was convinced that, in most cases that matter, a single approach to understanding is insufficient and that this principle has the widest possible application outside the sciences in general.

It is Simple, Whether One Likes It Or Not

The pursuit of simplicity, as I have tried to show in this chapter, is paramount in developing scientific ideas. Despite the initial complexities of quantum mechanics, the unification of chemistry and physics represents a great simplification. The reader may have doubts, particularly since quantum mechanics and classical mechanics have not been reconciled. I believe this lack of reconciliation to be of the essence of nature, and hence not a failure at all. I believe that this gulf is part of our universe, and that one must try to understand it

[17] I had a small but striking illumination on the peculiar nature of the theory myself. A few years ago during a visit to Athens, I observed a sign on a funicular: 50 Ατομα. I was greatly surprised that the vehicle was described as composed of only fifty atoms. I was told that a more correct translation would be fifty individuals—the limit the vehicle could carry. In 1912 Bohr already had an idea that there is a commonality between individual atoms and individual people. At least one can be mentioned—both change radically when divided.

fully despite its unusual aspects. Otherwise many modern developments will remain totally strange.

I have a final story for this chapter, offered as a little comfort to the reader who has struggled through new and perplexing ideas. One should never feel alone in this. As a young student beginning to "understand" these ideas, I had a marvelous proposition: teach the mathematics of operators[18] in elementary school—the rest of mathematics is superfluous to an understanding of microscopic physics. When I first had a chance to talk to Niels Bohr at a weekly tea for his collaborators, I propounded my idea at some length. During my presentation, Bohr slowly and quietly closed his eyes. I attempted to summarize as rapidly as possible and then waited through several eternities of silence for his comment.

At last Bohr, without opening his eyes said in a barely audible voice, "You might as well say that we are not sitting here drinking tea at all, but just dreaming it." Over the years, I have come to imagine a similarity between Bohr's comment and one of my favorite passages from Shakespeare:

> be cheerful, sir.
> Our revels now are ended. These our actors
> As I foretold you, were all spirits, and
> Are melted into air, into thin air;
> And, like the baseless fabric of this vision,
> The cloud-capp'd towers, the gorgeous palaces,
> The solemn temples, the great globe itself,
> Yea, all which it inherit, shall dissolve;
> And, like this insubstantial pageant faded,
> Leave not a rack behind. We are such stuff
> As dreams are made on; and our little life
> Is rounded with a sleep.[19]

Both Shakespeare and Bohr knew the ambiguous, even paradoxical, nature of that which we accept as reality.

[18] See Appendix XI.

[19] *The Tempest.* Act IV, Scene 1.

Chapter Four

SIMPLICITY AND THE THOUGHT PROCESS

As this chapter will demonstrate, considering simplicity together with the thought process of humans is almost a contradiction in terms. However, mankind has obviously pursued simplicity in a productive way through scientific understanding, and it seems sensible to consider some of the tools that have served in the pursuit: mathematics, computers and the human thought process.

If one looks at both mathematics and computing machines, one faces a very peculiar contradiction. Both computers and mathematics have made possible many advances in science and technology, but if one tries to describe what a mathematician is doing, one often ends up describing something rather simple and boring, as simple and boring as 2 + 2 = 4.

Computing machines are seen in the same way. The elementary operations that they perform are simple arithmetic. Worse still, they do it by counting on "two fingers." They don't seem to do anything except more and more of the same operation. Since computers do these calculations in principle at least a million times faster than humans, they do much more of the same thing and must seem even duller. How anything new or marvelous can come out of repeating exercises of this kind escapes most people. Is the equation "boredom + boredom = greater boredom" correct? Can anything of interest result from calculations?

Unexpected Simplicity: A Prime Example

I am going to discuss four examples that have come out of working (or playing) with mathematics. The first one happens to be a difficult mathematical theorem which contains an unsuspected regularity.

A prime number is a number that cannot be divided by any other integer[1] greater than 1 without leaving a remainder. Apart from 2, all prime numbers are clearly odd numbers. The list of additional prime numbers in ascending order is 3, 5, 7, 11, 13, 17, 19, 23, 29, 31, 37, 41, 43. . . . An interesting fact about these prime numbers is that all of them can be written in one of two forms—as four times an integer (**n**) minus one, ($4n - 1$), or as four times an integer plus one, ($4n + 1$). For example, $3 = (4 \cdot 1) - 1$; $5 = (4 \cdot 1) + 1$; $7 = (4 \cdot 2) - 1$; $11 = (4 \cdot 3) - 1$; $13 = (4 \cdot 3) + 1$; $17 = (4 \cdot 4) + 1$; and $41 = (4 \cdot 10) + 1$.

Can a prime number be written as the sum of two squares? At first, there appears to be no reason to ask this question. However, more than two hundred years ago, a remarkable French mathematician, Fermat, found an interesting theorem: a prime number of the form of $4n - 1$ can never be written as the sum of two squares; a prime number of the form $4n + 1$ can always be written in one and only one way as the sum of two squares. (The first statement of the theorem is easy to prove; the second, hard to prove.)

The number 13, from my examples above, can be written as the sum of $2^2 + 3^2$. The number 5 which also has the form of $4n + 1$ can also be written as the sum of the square of 2 plus the square of 1. It is remarkable that numbers should behave this way consistently. It is not hard to check the accuracy of this theorem through all the prime numbers up to one thousand, but to prove it for all prime numbers, including the infinite number of unknown primes, is a different matter. (See Appendix XII.)

A second example is another of Fermat's theorems, called "Fermat's great theorem." The question involved is whether the equation[2] $x^n + y^n = z^n$, can be solved with integer numbers for **x**, **y**, **z** and **n**. An example of a solution of the equation is $3^2 + 4^2 = 5^2$, which is often a schoolroom solution to the Pythagorean theorem. When **n** equals 2, the equation can be satisfied. However, this theorem states that, for **n** equal to 3 or more, the theorem can NEVER be satisfied using integers.

[1] An integer is a whole number, that is, not a fraction. Integers can be positive or negative, and are those numbers with which a non-mathematician is most familiar, such as 1, 5, 112, -4, -37.

[2] The expression x^n (n represents an integer) signifies that the number **x** should be multiplied by itself **n** times.

The theorem should be restated: the sum of two different integers, each raised to the same exponent, cannot be equal to a third integer raised to the same exponent when the exponent is greater than two; in mathematical shorthand,

$$x^n + y^n \neq z^n, \text{ if } n > 2.$$

This theorem has been proved for the case of $n = 3$, and for quite a few other n-values. However, Fermat wrote a note (in French) in the margin of a manuscript: "I have discovered a truly remarkable proof, which this margin is too small to contain." Then, before he apparently ever got around to putting the proof down anywhere, he died. Nobody has found the proof yet. This is particularly frustrating since almost all Fermat's other statements about theorems have turned out to be correct.[3]

There is a story about Fermat's great theorem which involves a prominent British mathematician, Hardy, who had a horror of airplanes. When he was in his seventies, he got an invitation to come to the United States to give a few lectures. He couldn't manage to come except by airplane. The conditions of the trip otherwise were so nice that he accepted in spite of his fears. Just before he left, he called a friend and said, "It's marvelous. I have the proof of Fermat's big theorem. However, I am just about to leave for the United States. I'll have to tell you when I come back."

Hardy went to the United States. He flew home. Three weeks passed. His friend heard nothing, so finally he called. "What about Fermat's big theorem and your proof?" Hardy responded nonchalantly: "Oh, that. I don't have a proof. I just hoped that God would not perform the same terrible trick twice." Hardy was not a particularly religious man and seems to have had a somewhat original way to take out "life insurance."

A third theorem, another that every young mathematician hoped to prove or disprove, is called the four-color theorem. I have a special reason for including it which will be clear at the end of this chapter. In this theorem, one must have a set of contiguous shapes in the same two-dimensional surface, all touching one another—like the 48 states of the United States, with two exceptions: no state should be divided like Michigan, and patterns like the "four corners" should be excluded. (Only three shapes should meet at any point.) Using only four colors, is it possible to color this pattern in such a way that, wherever two shapes touch, the colors are different? It is easy to prove that it can't be done with three colors. However, can it be done with four? Try it on Figure 4.1. Five colors are obviously sufficient,

[3] Only one of his suggested theorems has been found wrong. Proving it wrong was not an easy trick.

Figure 4.1. A Set of Continuous Shapes for the Four Color Theorem.

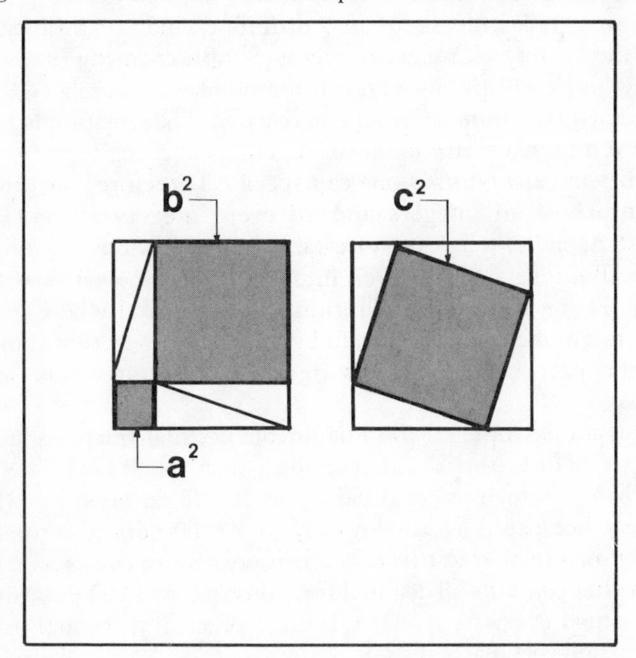

Figure 4.2. An Alternate Proof for the Pythagorean Theorem.

and in this case, four colors are enough, too. But whether or not four colors are sufficient in all cases has been debated for a very, very long time.

Since Pythagoras has been so much included in the book up to this point, he shall not be excluded from the chapter on mathematics. The four-color theorem includes a hint of geometry, the mathematics of space, so this seems an appropriate time to introduce a proof of the Pythagorean theorem which was not the method that Pythagoras used.

Consider two large equal squares as shown in Figure 4.2. Subtract from each the same four equal right triangles. The result of the subtraction, shown as the square shaded areas, must be equal. What is left over from the first square is a^2 and b^2; what is left over from the second is c^2. The equation $a^2 + b^2 = c^2$ is the obvious consequence.

Is Infinity Equal To Infinity?

While still discussing mathematics, I shall turn from theorems to concepts. Most people believe that infinity is infinity—that there is just one kind of infinity. Actually mathematicians have proved that there are an infinite number of infinities. This is not very easy to show. I shall talk only about two different kinds of infinity.

Take the infinity of integer numbers. Simple counting (if one lives infinitely long) will produce an infinite number of numbers. Sooner or later, any given number would be reached. The situation is similar if only even numbers are mentioned. The progression goes on into infinity beyond any number one can specify. Therefore, one can state the infinities of all integers and all even integers are not really different. Actually, it is easy to establish a one-to-one relationship between all integers and all even integers: $1 \leftrightarrow 2$, $2 \leftrightarrow 4$, $3 \leftrightarrow 6$, . . ., $n \leftrightarrow 2n$. There are, however, infinities of the kind where it is not possible to enumerate a set of numbers in such a way that sooner or later some particular number within the set of numbers will be reached.

Take as an example all the imaginable decimal fractions starting with zero before the decimal point, such as 0.123123. . . or 0.314159. . ., which is π divided by 10. π can not be written completely because it is known only to 500,000 digits, although it may be known further at this time.[4] Is it possible to construct a list of numbers that contains all decimal fractions that proceed to infinity—a list in which every particular infinitely-proceding decimal fraction will ultimately be reached? *No, it is not.*

[4] π is the ratio of the area of a circle divided by the square of its radius.

To prove this fact, assume first that it is possible to do so and that there is a list compiled of such decimal fractions. Using that list, I will find an infinite decimal fraction that is not contained in the list. For the first decimal, I shall choose a number that differs from the first decimal in the first fraction. For the second decimal, I choose one different from the second decimal in the second fraction on the list. The third decimal I choose is different from the third decimal in the third in the list of fractions, and so I proceed down the list. I am moving on the diagonal, so the number I produce will always differ at least in one decimal from any number given in the list.

Thus my originally proposed list is necessarily incomplete. I conclude that no complete list can ever be constructed. This kind of infinity is called, for obvious reasons, a non-denumerable infinity—a most peculiar concept. It is possible to go on from here and construct still other different kinds of infinities, a fact discovered less than 100 years ago. (Even mathematicians can think of new things.)

No sooner did mathematicians start to play with infinities than they started contradicting themselves, which, of course, created a big scandal. A controversy developed as to whether all the mathematical questions that could be asked could be answered, at least in principle. For example, when π is written in decimal form, one can show that it has an infinite number of decimal figures that follow in some apparently jumbled fashion. π starts with 3. Does 3 occur again? The answer is yes. Is there a final number 3 in the decimal fraction beyond which 3 never occurs again? Nobody has even the faintest idea of how to begin searching for the answer. Can the question of whether there are a finite number of 3's in π be answered at all?

In 1930 Goedel, an excellent Austrian mathematician, proved something very peculiar. In each branch of mathematics as complicated as real numbers, one can construct a proposition which cannot be decided in that system of mathematics. In such a case, one can postulate that the proposition is true or that it is false. In either case, one can build a new system of more complex mathematics on the proposition. This being true, mathematics will never be complete. It will have no end and will never become a closed and finished science. Mathematics itself has become infinite.

A Most Valuable Ally: The Computer

I am now going to change the subject to one that mathematicians originally found very uninteresting—I am going to talk about computing machines. Their practical development started with

International Business Machines, or IBM as it is known today. Computers started with the cash register, with a problem no more exciting than bookkeeping. Bookkeeping is, however, an enterprise where it is preferable not to make mistakes.

In the late 1930's, I met an astronomer named Schwarzschild whose father had lived in Austria and had a lot to do with relativity. The young Schwarzschild (who is now the old Schwarzschild) was using the original IBM machines to calculate the orbit of the moon. Newton also calculated that orbit—in an incomplete manner. Why, one may ask, is it so difficult to do? If only two bodies are considered, the calculation is simple, but taking three bodies into account makes the problem extremely complex. The orbit of the moon is influenced by the sun, so three bodies—the earth, moon and sun—must be considered.

What is too difficult for mathematicians can be done by the computer. Schwarzschild calculated the position of the earth relative to the moon, and the position of the moon relative to the sun, and all the interacting influences from moment to moment. Since the calculations had to be made in very small steps, this was a very slow and repetitious project. The principle was not so difficult, but the calculations were painstaking. Today problems of this kind are routinely solved by computers.

At the same time, these early computers were applied to another type of problem in England. I believe this application had more to do with winning the Second World War than either radar, the atomic bomb or liberty ships. British and Polish spies had learned the general design of a machine that the Germans were using for transmitting coded messages. Such secret messages must be able to be encoded and decoded quickly, so there had to be a system within the code.

Because the British knew from its design roughly how the machine worked, they could narrow down the possible codes which it could use. Early computing machines then made it possible for them to try out many possible codes and discover the one which made sense. The Nazi code machine had been nicknamed "Enigma," and the British broke the Enigma code with the help of computers. Even though the code was frequently changed by the Nazis, the British computing machines could keep up with the changes, a feat that unaided human mathematicians could not have accomplished. Throughout the entire war, the Allies listened in on the Nazis' conversations. This was really decisive. The British fliers could not have defeated the much stronger German Air Force, even with radar, without the knowledge of where and when the next attack would come.

Another example of how important knowing the code was involves the excellent German general, Rommel. He was rolling his tanks through Africa to take Egypt. The terrain on Rommel's right flank in the south, the Qattara Depression, was extremely rough. Rommel's plan was to surprise the British forces by sending the attack through the Depression, hitting the British forces from the flank. Not only did the British know that this was the plan, but they even knew how many tanks were coming and how much ammunition was needed to stop them. This was an enormous advantage and certainly changed the course of the battle.

The Germans also developed V-1 and V-2 rockets which would have been extremely effective (and indeed were—when they were available to be used). The rockets might well have prevented the Allied invasion of France. However, the Allies learned where these rockets were being made, again, because of breaking the Enigma code. They were able to bomb the rocket factories and limit the supply of rockets sufficiently so that the invasion was not affected seriously.

In the United States, during this time, computing machines served an additional purpose. In order to construct an atomic bomb, a number of extremely difficult calculations had to be made. They were too difficult for us to perform by hand. Johnny von Neumann came to help with the work in Los Alamos and showed us how to use the computing machines. They were extremely clumsy and could hardly work faster than a person. However, they could work more persistently and practically without error. They were slow because they relied on mechanical electrical relays and were "programmed" with wiring and switches. What the computer would do next depended on the results already obtained and on the position of the switches. The switches could be left unchanged most of the time, but when they had to be adjusted, they had to be moved by hand.

Today computers have become so immensely important that more than one million people are trained as programmers for computers. This never would have happened without great advances in design. Thinking machines—computing machines—were introduced by a variety of people, but the greatest changes in them were made by mathematicians, particularly the very outstanding mathematician just mentioned, Johnny von Neumann. Von Neumann persuaded IBM that entirely new methods were possible. Instead of hand operated mechanical relays, electronic components were built into the machine.

The first electronic systems were similar in construction to the old radios, built with vacuum tubes. The usefulness of such tubes is that they can be switched electrically from open to shut and back again. Within the vacuum, electricity flows from one wire to another. Between the two wires there is a wire gate. The gate between the wires can be charged so that the current is either allowed to flow or is stopped. The result is that with a small input to the gate, a big output can be produced throughout the machine. The system of currents flowing through complicated electric circuits can be switched on and off at least a million times faster than electric currents can move mechanical relays.

Later, small systems of solid material were substituted for the vacuum tubes. Semiconductors—substances that have not quite made up their minds whether to be conductors of electricity or insulators—are now used instead of vacuum tubes. The similarity between vacuum tubes and silicon discs is not apparent, but the similarity involves an idea presented earlier in connection with Niels Bohr's work. He pointed out that the electrons in atoms can have only certain sharply defined energy values. All other energy values do not fit into the atoms.

In a solid, a whole range (or band) of energy values can be occupied, while other bands must remain empty. In a metal, there are enough electrons to fill only a part of a band, and therefore electrons can accept very small amounts of additional energy. In other words, the electrons in a metal are highly mobile (conduct electricity well). In an insulator, a complete band is filled, and electrons can accept only big additional chunks of energy.

In some semiconductors, the gap between the bands is quite narrow. In others, there may be very thin bands produced by impurities (that is, different atoms with different energies) introduced in the proportion of a few parts in a million. In these cases, the presence or absence of relatively few electrons can make a great difference in conductivity.

The similarity between a vacuum and a crystal of a semiconductor is that the electrons due to their particle-wave nature may move in both practically unhindered. In a solid, this is possible only when electrons have appropriate bands to enter. Due to an extremely important principle, the Pauli principle,[5] all electrons must have different states, and so the conduction can be strongly affected by just a few electrons in an otherwise empty band, or the absence of just a few electrons in an otherwise full band.

[5] In the absence of the Pauli principle, heavy atoms (which have strong charges on the nuclei) would be very small, due to the attraction of all the electrons into the orbits closest to the nucleus. The principle states that once the low states are occupied, higher electron states must be filled in. The result is that heavy atoms are, in general, somewhat bigger than light atoms.

The presence or absence of these few electrons can play a similar role in permitting currents to flow just as the charge on a gate in a vacuum tube does. The practical point is that what can happen in a vacuum tube in a big volume can be reproduced in a semiconductor in a much smaller one. Semiconductors can be changed rapidly from conductors to insulators. This on-and-off situation is all that is needed to perform simple operations like counting or adding, or to place a new number in the "memory" of the machine, or to recall something from the memory. The procedures have increased in complexity, but all are performed on the same simple basis. The main point about semiconductors is that they require much less space than vacuum tubes do. They require less energy to drive them, and they are becoming steadily less expensive.

I wear an electronic watch that calculates the date, operates as a stopwatch, or as an alarm clock, and also calculates the time. It is a smart little computing machine. If my watch were made with vacuum tubes, it would be larger than a grandfather clock, and I certainly couldn't carry it around. The development of semiconductors made it possible for people to wear computing machines on their wrists and carry calculators capable of really complicated work in their shirt pockets. Because little circuits can be etched on the surfaces of the semiconductor, a tremendous amount of computing power can be packed into a very small piece of apparatus.

Work on combining the most modern circuitry into huge and powerful computers is in its early stages. Big companies have been very slow in tackling the problems involved. One of my friends in the Lawrence Livermore Laboratory, Lowell Wood, is trying to integrate modern hand-computer circuitry into a really large machine. If he succeeds, he will improve the performance of the present large-sized computer one-hundredfold.

One of the most difficult things involved in developing a really powerful computer is to program it, to plan how the work which must be done should be organized. All of the choices must be foreseen; all the interdependencies of the operation must be identified. The bigger the machine, the more difficult this programming becomes. Lowell Wood and his co-workers are using a tremendous number of very cheap, very small components to create the most complicated machine ever built. It is so big that no one can program it, so they have designed a way for the machine to program itself. Only the general outline of the program is entered. The details are worked out by the machines.

There is an ever increasing number of uses for computers. During World War II, the military men analyzed the past patterns of submarine activity and entered these statistical data into computers. This made it possible to calculate probabilities as to what the best routes were in searching for submarines, what the best size for a convoy was, what tactics best allowed a ship to escape. This kind of analysis is called "operational research," and it worked more than thirty-five years ago. Today, businesses use this same procedure to keep track (with good results) of how much stock they have, how the stock should be distributed, when they should buy new stock, and when they should sell. One can also use computers to keep track of electrical networks and their interactions so that if a generator fails, others may or may not be called upon. Currently, operational research is used in many widely divergent settings and is proving very useful.

However, even good things can be abused. Using computers, people have tried to make plans about the future. What is fed into the computer is often a set of not very reliable assumptions. The result of this planning is aptly described as "garbage in—garbage out." The worst consequence is that once a computer has spoken, the customer believes it. In Herman Kahn's words, "garbage in—gospel out."

A more important application of computers is the calculation of complicated physical phenomena. Today, weather observations on a global scale can be obtained through the use of satellites. The winds and clouds of the planet are coming to be known at every moment and altitude. However, a great deal more information is still missing, but it could eventually be obtained. If this happens, the computing machines will be able to predict the weather reliably, maybe a whole week ahead. This is a most practical application of computers.

Computers can also be efficiently used to calculate the air flow around the wings of newly designed aircraft. This air flow is very complicated because it is turbulent. Today, before an airplane flies, a model of it is constructed on a miniature scale, and the model is tested in a wind tunnel to see how it will behave in actual flight. Computers will eventually do a better job. Combining such calculations with experiments in wind tunnels may give near perfect assessments.

Is the Human Brain Obsolete?

Having given so many examples of the superior abilities of thinking machines, I would like to consider whether our own brains should be considered inferior kinds of computers. One could hardly carry the idea of the dehumanizing effect of technology farther than to say that computing machines have made the human brain obsolete. Therefore, I want to discuss this problem in a rather detailed fashion.

First of all, there is no question that the computing machine can perform the same function faster and more reliably than the human brain can perform the same functions. In speed, reliability and accuracy, the computing machine is vastly superior to any brain. There is, however, one very important restriction. The functions of a computing machine are limited to those thinking operations which people can describe in a precise manner. Before one can make a computing machine do work, one must know exactly how the human brain does that work (or at least some way to get that work done).

In addition, there are incredible numbers of everyday human tasks accomplished with ease, even by individuals whose mentality is below average, that the computer cannot begin to do. People recognize individual people when they meet them or see photographs of them. A machine is unable to recognize people. There is not even a good start at giving a machine that capability.

Translation from one language to another by computer is also beset by immense problems. It is said that somebody once asked the computing machine to translate the phrase, "out of sight, out of mind," into Chinese. The computing machine promptly sent back a set of characters. Then it was suggested that the characters be translated back into English, so this request was made of the machine. Back came "invisible idiot." Indeed—out of sight, out of mind.

The meaning of words is determined by usage, by context, by associations. Someday, it may just barely be possible to put all of these factors into the computing machine. At present, because the language of mathematics does not depend on context and associations (it is truly a much simpler language), the computer is much more effective in solving problems dealing with this terminology.

I have a favorite pipedream, and I wish a machine would be programmed to turn it into reality. I would like to have a computer that would make jokes. However, I expect a few problems in designing this machine, because while I can identify a joke when confronted by one, I don't know how to define a joke. There are many published discussions of humor, but there is no general agreement on the psychology of a joke. In one case, that of writing limericks, at least the formal rules are known. It is possible that information about humor could be gained by testing theories about it with a computer.

One could design a computer to write limericks. The test of the success of the theory would be to publish a book of limericks, half man-made, half machine-made. If the reader couldn't tell them apart, then the theory of what a joke is would have progressed.[6] The computer can be used in this indirect fashion to understand more about ourselves and our brains.

In fact, a new branch of psychology which is a study of the processes of knowledge, cognitive psychology, has come into being largely because of computers. The use of computers for exploring the functions of the human brain raises a particularly interesting question. Should people consider their brains as a mechanical or as a biological instrument? Lest the answer to this question be given too quickly, I will rephrase it. Is the human brain essentially more like that of a dog or like an IBM machine?

I am afraid that my brain (and everyone else's) is more similar to a dog's. This somewhat unflattering statement can be made in an even more objectionable manner. A dog knows nothing of logic. People know a little about logic but use logic painfully and sparingly. Indeed, I find a paradox in the nature of logic. (This may be due to my love of paradoxes which has been well developed in the study of physics.) I am tempted to say that logic in humans is a perversity. Of course, logic is also one of the most precious human gifts. To me, man's relationship to logic is an inherent paradox of human nature.[7]

A thinking machine is programmed; this means that each step it takes is rigorously prescribed. The instructions are no less precise whether one instructs the machine to choose a random number or instructs it to translate a word. The machine surpasses the human brain only insofar as the basic functions can be clearly defined and are able to be described as a specific sequence of steps.

Let me repeat this point by considering a word entered into or produced by the machine. A word is much more than a mere assembly of symbols.[8] Its meaning resides in its function. There must be a precise function assigned to a word by the plan which organizes the machine's operation. This function is a rigid prescription of how the word is to be used. This, in fact, is the meaning of a computing machine *program.*

[6] There was a professor named Teller
Whose physics (and humor?) were stellar.
He planned a computer
Whose jokes would be cuter.
Do you think it can write a best seller?
Anon.

[7] It is not an accident that *logic* is derived from the Greek word *logos* which means *word.* Logic has an advantage over paradox: people can talk about it more easily and clearly.

[8] Except in the case of word processing computers. There a word is only an assembly of symbols.

In contrast, consider the function of a word in human usage. The word is effective through the exceedingly numerous associations which are established between it and other words, for example, associations that come from context. When visiting in Athens, I saw the word Μεταφορα. The word *metaphor* sprang to my mind. But the word was printed on a moving van. While the spiritual Anglos use the word in terms of transferring a meaning from one description to another, the practical Greeks employ it to the moving of furniture. The connotations of words are clear and, in the majority of

"Now that we've got *this* wrapped up, I'd like to get into math."
© 1977 by Sidney Harris, from the *American Scientist.*

cases, shared by all native speakers[9] of a language. There are even experiments which suggest that connotations of words are perceived before the actual word is recognized. The function of a biological word is not rigidly determined (as far as anyone knows) by any predetermined program. If, indeed, such a program exists, it is vastly more complicated than can be guessed at present.

The biological brain, even a comparatively simple one, is endowed with an exceedingly large memory which makes many associations possible. The comparatively limited memory of a computer has so far frustrated machine attempts to recognize a handwritten letter (alphabetic symbol), let alone to recognize a face. The major problem is to identify the differences between the mechanical brain and the biological brain. If a bridge could be built to connect them, then the computing machine might be able to accomplish a wider variety of tasks.

In principle, it might be possible to reduce all biological functions to mechanical functions. It is also conceivable that by understanding parts of the biological function of the brain in a mechanical manner, one could better isolate and thereby illuminate those parts which are not as yet reducible to mechanical explanations. Indeed, there may be brain functions—functions of our own minds—which will never be understood on a mechanical basis. If so, the more these functions are isolated, the clearer people will become about a subject which is currently most obscure. Studies utilizing computing machines may make surprising contributions to the exploration of the human mind.

The Man-Machine Merger

In actuality, machines can take over what is routine and do it better than humans, provided humans already know precisely how to do it. In all other respects, a person's thinking ability is vastly better than a machine. But there are a great many situations in which a combination of thinking machine and human can perform a task or solve problems far better than either could alone. A computer assisted in the preparation of this manuscript. It did the cutting and pasting, and it even proofread for spelling errors.[10] It typed and retyped at an incredible speed and relieved the humans involved of the monotony of that task. It made it possible by its efficiency to get the manuscript to the publisher almost on time.

[9] The restriction to native speakers suggests how deeply the roots of association reach. I, for one, speak German like a Hungarian, English like a German, and Hungarian like an American.

[10] The reader will understand that I am not unwilling to pass the buck.

On a less practical level (but one which has always interested me), no machine has been constructed yet that could beat a real chess master. Although the rules of chess are clearly specified, the number of possibilities is too great for a machine to handle. Yet a chess master can cope with them. A thinking machine and a man together, a chess master with a computer, probably would easily beat any competitor who had no thinking machine, provided the chess champion could effectively communicate with the machine.

My statement that a man and a thinking machine together are better than either alone has been proved in connection with the four-color theorem that I discussed earlier. The theorem is correct—four colors are sufficient. The machine alone could not attack the problem, but the very best mathematicians could reduce the problem into a great number of routine operations. These routine operations were then carried out by the computer. In this way progress in mathematics was made possible by the man-machine combination. Please note again how easily the mathematical theorem was formulated and how desperately complicated it was to prove it.

I believe that what the machine does is the opposite of dehumanizing. It allows us to get rid of the routine. It forces us to concentrate on the kinds of things that the machine cannot do, and there will always be a great number more of these. Recall that I mentioned in mathematics as it is understood today, one can construct questions that cannot be answered within the system in which the question was formulated. No matter how the machine is constructed, there will always be situations that the machine cannot foresee. The human element, the element of invention, the element of true ingenuity and flexibility, is not apt ever to be taken over by machines.

Some of the developments I have considered in this chapter will not happen in my lifetime. They may not even occur in the lifespans of my youngest readers. But the man-machine combination is likely to transform human thinking and transform it into something much more intricate, very much more interesting and incomparably deeper than mankind has as yet encountered.

Chapter Five

SCIENCE AND SURVIVAL

Science and technology have brought about great changes in the way people live, changes which move people everywhere at an ever accelerating rate into an unknown future. So far, I have talked about science, but now it seems important to turn to the interaction of science with human affairs. We are faced with problems of a completely different kind here. The future is, of course, unknown. By contrast, the past is wrongly known. History, most often, is recorded with a great amount of undeniable and systematic distortion.

In the recorded ages of primitive technology, the human pyramid was necessarily built on a broad base of people who spent their lives in hard labor. Spiritual accomplishments could, as a rule, be achieved only near the top—by comparatively few people. That history should overemphasize these peak accomplishments is understandable and unavoidable. Not only are the masses of common men and women neglected, peak accomplishments are seen through a further filter: the eyes of historians. Most people, including historians, have a point of view that encourages them to place importance on particular ideas. Objectivity in describing other people's actions is best achieved if one abstains from claiming objectivity and instead frankly states what one believes, that is, openly proclaims one's prejudices. Of course, even then, objectivity remains approximate.

In making this statement, I clearly have the obligation to make my own prejudices clear. Perhaps I might summarize them by seeing our present age as a repetition of the story of the tower of Babel.

Man's great accomplishments may, indeed, set nation against nation and result, not in the annihilation of mankind, but in strife and unhappiness. Yet, it may also result in the elimination of degrading poverty. It may even permit us to shape our future and escape being mere toys of our natural surroundings—which have not always been kind to the human race.

Thus this chapter becomes a kaleidoscope of the bright and the dark, all of it related to the new knowledge that sprang from science. But first we shall look into past paradoxes. They may make future difficulties easier to accept.

The Mirages of a Single Perspective

As space is curved, so are the spectacles of historians. The bigger the mass (the more important the event), the greater the distortion due to the unavoidable curvature. Thinking about the most important events of the past, I am struck by the ambiguities in two truly outstanding cases. One is the revolution in religion, morality and art introduced by the Pharaoh Ikhanaton almost fourteen centuries before Christ. As far as we know, he was the first monotheist, the first to attempt to found a government more on love than violence, and undoubtedly the first whose realistic reforms in art survive to this day. Countless people, among them our contemporaries, have admired the sculpture showing his chief wife, Nefertiti.

Yet his accomplishments were wiped out by a counter-revolution, and he was accused of a great variety of horrible crimes. From a historical perspective, one has to admit that probably not all the accusations against him were false. It is better to admit that what may have been the single most wonderful reform in spiritual history cannot be properly evaluated on the historical record.

The second example is closely connected with technology. The great North-South Canal, constructed around the year 600 A.D. in China, was much more than an amazing engineering accomplishment. Its construction helped unify two widely differing regions of China and undoubtedly helped to relieve the famines in that country. It was constructed under the short-lived Sui dynasty, which is condemned in Chinese annals for having sacrificed many lives to the canal. The second (and last) emperor of that dynasty is described as having built the canal only to serve for his pleasure trips.

History may be a valuable teacher, but it would be a more valuable one if it would make more use of the concept of complementarity, a concept which I propose to employ repeatedly in this chapter. One conclusion appears inescapable. The general development of history is more reliably ascertained than the individual event, no matter how dramatic and significant it may appear.

Technology once introduced is rarely forgotten. The Chinese canal has been maligned, but it is still in service today. The methods of providing food, clothing, housing or weapons have all shown irreversible trends. Whether these trends have been for the better or worse is, in fact, debatable. But when one considers that an ever increasing number of human beings do live on this planet, and that a greater fraction of them are freed from the necessity of the severest kinds of labor, one can at least argue that the understanding and control of nature has been a positive contribution.

During the flowering of Greek civilization, which some people today consider a Golden Age, there was widespread suffering, the Persian, Peloponnesian and Macedonian wars, and an almost universally harsh existence. It is questionable that many of today's Greek scholars would enjoy being contemporaries of Plato, the unsuccessful advisor of a Sicilian tyrant. In Greek literature, one finds reference to "an original Golden Age," which degenerated into a Silver Age and finally into the ancient Greek period called the Iron Age.

The third of a century since the Second World War has been filled with remarkable events. In truth, it may be looked on as a period of marvelous progress in human well-being. In 1945, the number of people who led reasonably comfortable lives were hardly more than 200 million, most of them living in the United States. The third world (as it is called today) and all the countries ravaged by the war were truly in terrible condition. Less than 10 percent of mankind was not on the ragged edge of misery.

Today, in 1980, Europe, Japan and Russia have recovered. Misery is still the lot of most of the world, but the underprivileged comprise three-fourths rather than in excess of 90 percent. The change could never have been made without the aid of technology. What may be even more important is that in Europe, age-old enmities have abated. On the spiritual side, this is fully as important as the material improvement. But this good news goes unnoticed. This is a well-known property of the press in the free world. Improvements are never dramatic. Life improves slowly and goes wrong fast, and only catastrophe is clearly visible. In a historical perspective, there can be no doubt of the overall trend. The lot of considerable numbers of people improved, not only materially but also in moral respects. This is not disproved by the occurrence of catastrophe, but neither should the setbacks that catastrophes cause be belittled.

A Paradigm: Energy Shortage

From a historical perspective on technology, the main question is to what extent natural forces can be controlled and made to serve humans?[1] This development depends in part on the availability of energy. It is a remarkable fact that the precise concept and understanding of energy is a recent accomplishment of science. The beginning of the Industrial Revolution is usually dated from James Watt's invention, which was not a steam engine, but an efficient steam engine. Watt recognized efficiency when he saw it, but he could not have explained it. The circumstance that efficiency is the percentage of energy utilized and that energy itself is conserved was not clarified until seventy years later. When one talks today about the energy shortage, one means a shortage of usable energy, that is, energy that can be transformed into mechanical work.

Energy has played an important (but not exclusive) part in technical and social development. In the steps of lightening labor, mankind has exploited animals as a source of energy, exploited other people as that source, and finally is exploiting the energy sources of nature in machines. To complete this picture, one may add that the exploitation of other people provided the most intelligent (if not the most obedient) form of energy in the past. Modern machines coupled with the marvels of electronic computers deliver energy in highly intelligent and almost completely obedient form.

Today, the reality of energy shortage makes this a topic of lively discussion. One of the postulates accepted in the current discussion is that society should get away from exhaustible energy sources and concentrate on renewable forms.[2] As is true of many other widely accepted statements, this one is often narrowly interpreted and somewhat erroneously perceived. Solar energy is described as renewable, and, indeed, the source of radiant energy is apt to be constantly renewed for billions of years to come. But the means of exploiting solar energy, once constructed, will deteriorate and will have to be replaced. All plans for solar electricity result so far in quite expensive energy. This circumstance is a correct indication that one basic law of economics has a wide applicability: not only is it true that there is no free lunch, but also, there is no free solar energy.[3] The continuing

[1] From the moral point of view the question is, of course, whether control in human hands is used or misused. It is important to acknowledge that moral judgments are not unanimous.

[2] In the long run, one can hardly disagree.

[3] Solar energy in agriculture (biomass) may be considered free. It is nevertheless unavoidably coupled with human labor, without which this energy would be wasted.

care and reconstruction of mirrors, boilers, solar cells and storage equipment will consume both exhaustible materials and the kind of labor which is limited. One may, of course, argue that human labor is renewable (though not inexhaustible). One should not forget, however, that the main point in technology is to relieve the burden from those at the bottom of the human pyramid.

But perhaps hydroelectric power can be considered inexhaustible. Certainly, the construction of a dam will require both labor and energy, but after that, the dam continues to deliver power as long as the lake behind the dam is not filled with silt. Once that happens—which may take centuries, though sometimes less—new and generally poorer dam sites will have to be selected.

While these two examples can be considered in some approximation as renewable energy sources, a third one often mentioned, geo-thermal energy, can certainly not be so classified. Enormous amounts of energy are stored near the center of the earth. They trickle to the surface in such a slow manner that the effect they could exercise on total energy consumption is negligible. Practical geothermal energy sources can be described as fossilized volcanic resources (where hot lava has intruded or erupted, one frequently finds regions of exceptionally hot rock near the surface). There the energy can be and is profitably used. The total energy available in this form is hardly greater than the total that could be obtained from coal. At the same time, these geothermal sources have been deposited in millions of years and could be exhausted in a comparatively short time.

Should one therefore give up hope of finding essentially inexhaustible and practical sources of energy? No. A first strong reason for optimism is nuclear energy, which is a scientific and technical novelty of the last half-century and which, contrary to widespread fears, turns out to be safe, clean and, on the whole, less expensive than other energy sources. To produce agreement on such a vigorously debated subject in a few lines is impossible. I therefore merely recommend to the doubting reader that s/he should try to enumerate any victims of nuclear reactors and compare this number with those killed in coal mine accidents, to mention only one of many dangers.

Continuing present practices, uranium resources will serve for a finite period, which may not be longer than the period in which fossil fuels will be exhausted. This is due to the circumstance that only a very simple and obvious design of nuclear reactors is being used. With this design, only one percent of the energy which is in principle available in uranium is utilized. One could use all the energy available in uranium and in a similar more abundant element, thorium, and also the energy available in other nuclear resources which rely on fusion rather than fission.[4] The question of whether these energy sources can be utilized is a question of time and human ingenuity. If they are utilized in a more thorough fashion, the amount of fuel necessary will become small enough that it can be extracted from extremely widespread and abundant sources. Due to the high energy-content of the fuel, it will pay to extract it even when the fuel is present in small concentration. In this sense, nuclear energy could provide mankind with a source of energy sufficient for many thousands of years.

That one should demand virtually inexhaustible energy sources is justified. However, the only such source that I know of is human ingenuity. It is worth noting that there are no natural resources until humans discover a use for a particular substance. Until then, the substance is merely a part of the scenery. The same is true not only in the case of energy sources but also for other technical aids which certainly can influence our future. For this reason, I would call discussion of energy resources a paradigm.

Fear, Hope, Impatience

One may view the future with fear or with hope or with frustration and impatience. It may not be too great an over-simplification to say that the dominant sentiment in the Western world is fear, in totalitarian regimes in technologically developed countries it is hope, and in the third world it is impatience. This seems like a very strange situation. Is it true? and if so, why?

People—in fact all living creatures—seem to be inherently conservative when the situation appears to involve their personal well-being. Natural instincts have full and untrammeled play in a free society. Once fear and uncertainty occur, they can be contagious, particularly in periods of rapid change. Fortunately, the opposite is also true. History holds examples of periods when the prevalent sentiment was a dominating and soaring confidence in the activities of the time; the medieval cathedral building in Western Europe and the Renaissance as a whole are both good examples.

On the other hand, totalitarian governments cannot survive unless the people believe that life will improve. Once this hope is gone, stability is in danger. In a dictatorship, the luxury of speaking out, even the freedom of thinking, is restricted as far as possible to the leaders of the state. In this sense, the existence of fear in a democracy

[4] In fission, heavy nuclei are split and release energy. In fusion, light nuclei unite and release energy. Nuclei of minimum energy are the middle weight elements.

and hope in a dictatorship is by no means a praise of the totalitarian way of life. Rather it is a recognition of the fact that dictatorships must suppress doubt or they fall, while in a democratic state, freedom permits criticism and cannot forbid the fear of uncertainty.[5] Unfortunately, this situation is a major factor in creating the dilemma the people of free nations face when they try to look at the future in which unpredictable technology will play an increasing role.

Lest one think that technology has no deplorable consequences, one need only point to the newly mushrooming huge slums in the third world. London, Tokyo and New York are no longer singular in their size. Bombay, Calcutta, Djarkarta, Mexico City, Rio de Janeiro and a number of other cities in the third world have joined the list. Among the newcomers, one can find some of the worst poverty experienced by any people. That billions of people are caught up in rapid change is probably the most important event in recent generations. One should understand that the mood in the third world is quite different than that in the East or West. It is neither fear nor hope. Rather it is an impatience and frustration that change has not come fast enough.

The fear that plagues many people in the free world today has been discussed rather extensively.[6] A best seller a few years ago, *Future Shock,* certainly has a provocative title. Reading it, I discovered to my amazement that the shock of a housewife who has to choose between a great many goods in a supermarket is all but debilitating. Truly, this is a parody of a shock. Technology offers people, in the main, those things that are desired. The unfortunate hesitating housewife may have been spoiled by the good life in which she was rarely exposed to the shock of empty shelves or the despair of bread lines.

When one observes a child growing up in this new world, one can see another kind of shock, a shock that comes from the violation of an ancient human understanding. Children of abundance have little understanding of work. They lack a sense of the necessity of work for their existence and for the continuance of their society. In addition, many of these children have known little physical pain and so have a distorted view of happiness. Happiness for them becomes the "obvious" human condition.

[5] Expatriates of the Soviet Union and its satellites have a series of jokes which they share. Here is one from Budapest. A young, efficient woman obtained permission to leave for the U.S. Her boss was most distressed to lose her, and he asked, "Is your salary too low?" Said she, "I can't complain." "Are your hours too long?" "I really can't complain." "Are you unhappy with your housing?" "I can't complain at all." "Well, why are you leaving then?" "I'm going where I can complain."

[6] Half a century ago, Roosevelt said, "The only thing we have to fear is fear itself." The fear he pointed out was overcome in his time. However, his statement retains a great deal of validity.

But this same generation has also had to adjust to some difficult problems, problems which are real and dangerous but which are not immediately visible and can be repressed. Before 1945, Americans enjoyed an isolation guaranteed by ocean barriers as well as by the fact that the United States played a dominant role in the Western hemisphere that was beyond doubt or challenge. The bomb of Hiroshima ended American isolation. All people have become inextricably part of the human brotherhood, a brotherhood not exclusively characterized by brotherly love.

The Myth of the Apocalypse

Mankind's greatest collective fear, the doomsday fear, currently seems to have fascinated an incredible variety and number of people. Prophecies of doom are nothing new. Some of mankind's oldest myths have to do with the destruction of the world by fire; perhaps this is related to the first surprising and dangerous use of technology, the use of fire. Today much of the doomsday thinking centers on nuclear war. The modern myth is "a full scale nuclear war would be the end of mankind; at the very least, such a war would wipe out civilization."

It seems appropriate, therefore, to review a report issued by the National Academy of Science in 1975: *Long-Term Worldwide Effects of Multiple Nuclear-Weapons Detonations.* The title of this report means (in plain language) the consequences of an all-out nuclear war. It does not discuss the effects on the combatants; those effects are assumed to be catastrophic. Rather it considers the effect of a nuclear war on the neutral nations not involved in the hostilities. The most remarkable circumstance about the report is that its results have remained practically unknown. The conclusion is that, as far as one can predict, the uninvolved nations will not suffer serious damage.

In transmitting the report, the Academy president, Philip Handler, warns that this is only one view—that of the people who wrote the report. Many uncertainties remain even after a thorough study is made. To assess the report, it is necessary to note what has been taken into account and what has and has not been found out.

First of all, the report ignores all defensive measures that it is possible to take against the effects of nuclear weapons. Not only potential combatants but also neutral nations can make reasonable and probably effective efforts to protect their citizens. Thus, an important safety factor has been neglected.[7] The report also assumes that in a nuclear war, ten thousand megatons of nuclear explosives—

equivalent to ten billion tons of TNT—will be exploded. Horrible as this figure is, it is not the upper limit of what could be delivered. It may well be the upper limit required by any concrete military plan.

Before discussing the imaginary situation of the report, I want to introduce some information based on actual observations. When confronted with a tragedy that has the magnitude of the bombing of Hiroshima and Nagasaki, one's inclination is to disregard the details of the occurrence. Most people are unaware of many relevant facts. First, in Hiroshima and Nagasaki, three-fourths of the population within 5,000 meters (3.1 miles) of the bomb's point of detonation survived. These unfortunate people had no knowledge of even the simplest ways to protect themselves. Yet they survived.

It is also instructive to consider the biological damage to the survivors[8] which is now visible, thirty-five years after their exposure. Of the 300,000 survivors, about 80,000 were included as a representative group in a lifelong study of the effects of radiation on their health. There were 2,700 survivors in this study who had radiation doses in excess of 200 roentgens, and some may even have received 600 roentgens.[9]

The genetic effects of radiation that have been observed thus far are that there is no statistically significant increase in inherited birth defects. The popular misconception that future generations have been doomed by exposure to the bomb blast is probably a result of the fact that in the nine months following the bombing, there was an increase in the number of children born with birth defects. It is a well known fact that fetuses are extremely sensitive to radiation as well as to other damaging agents—if a pregnant woman consumes an excess of alcohol, the incidence of birth defects rises. However, in the survivor's children conceived after the blast and in their grand-

[7] Historical differences between the United States and the Soviet Union have no doubt played a part in determining these two countries' widely differing approaches to civil defense. Russia has been repeatedly visited by wars—most of them not of its own making. The youthful U.S., physically separated from powerful nations, has not had a comparable experience. The Soviets are prepared to evacuate cities, have built strong shelters for the populations that would remain behind and have well-distributed emergency food supplies. The U.S. is totally unprepared.

[8] The survivors have been nicknamed Ichiban by the researchers who have studied the health effects of the bombings. The word means "first people," and refers not only to the fact that the survivors were the first to experience the effects of a nuclear weapon, but also honors a particular attitude which these people possess. The survivors have participated in the study at their own time and expense over all these years in order that the effects of radiation may be better understood. (Data on cancer comes from a 1974 report.)

[9] Part of the difficulty that laypeople have in considering the effects of radiation is due to the various measurements of radiation. A roentgen is the designation of an arbitrarily introduced basic dose. A milliroentgen is a dose which is a thousand times smaller. At the Three Mile Island accident, the radioactive material released gave rise to a maximum exposure outside the plant of about 50 milliroentgens.

children and, in a few cases now, in their great-grandchildren, there is no identifiable increase in birth defects.

In studies conducted to determine the increased incidence of cancer, one form, leukemia, shows a significant increase in the survivors. Among the 80,000 survivors studied, there was *a total of 126* cases of leukemia. This represents more cases of leukemia than would be found in an unexposed population, but it obviously is not a frequent consequence of exposure to radiation. The other forms of cancer increased far less than leukemia. In the total population of these survivors, the total increase in cancer (including leukemia) is less than two-tenths of one percent.

I do not mention these facts to minimize the horrors of war, nor to suggest than an increase in birth defects or cancer from any cause is not a tragedy, nor that such a catastrophe as this bombing is not profoundly damaging to all mankind. I offer these facts in hopes that the physical realities of the situation will be more clearly understood.

Returning to the report, there seems to be little danger from radiation to people in neutral countries. The immediate effects are estimated at an average of four roentgen units per person in the northern hemisphere and at one-third this amount in the southern hemisphere. Radiation of this magnitude has not led to observable damage. One should not understand this optimistic-sounding statement as the complete absence of any damage. In an unprepared combatant nation more than half the people would probably die. However, a neutral unprepared country would suffer damage below a level that would have wide-ranging effect. The Academy report estimates that in neutral northern hemisphere nations, the incidence of genetic defects in live-born infants would increase over the current 6 percent level to somewhere between 6.01 and 6.1 percent, depending on the choice among available, credible theories.

The most remarkable fact about the Academy report is that the main worry has shifted from the effects of radioactivity to an entirely new phenomenon—a diminishing of the ozone (O_3 rather than the ordinary O_2) in the stratosphere and a small increase of ozone in the troposphere.[10] Ozone is generated by the absorption of high-frequency, energy-rich ultraviolet light from the sun by oxygen molecules in the atmosphere. Ozone itself absorbs "near-ultraviolet" light, light of somewhat less energy and frequency than "far-ultraviolet." Atomic explosions generate nitrous oxide (NO) molecules from oxygen and nitrogen. Nitrous oxide continues to destroy

[10] The stratosphere is a very high layer of the earth's atmosphere; the troposphere is lower and is the region kept in continuingly mixed state by heat convection and similar processes. Currently there is some ozone present in the troposphere, and when ozone is depleted in the stratosphere, somewhat more of it is generated by natural processes in the troposphere.

ozone. If there is less ozone surrounding the earth, more ultraviolet light (which produces sunburn and skin cancer) reaches the earth. Ozone is most important in the stratosphere, and it would be suppressed there by multiple nuclear explosions which would deposit radioactivity at very high altitudes. The results are somewhat uncertain but roughly predictable.

These effects would occur due to individual nuclear explosions of about a million tons of TNT or more. For at least some time, the number of ozone molecules might decrease to roughly one-half. This would lead to an approximate increase of 10 percent in skin cancer. It also might influence the growth of some essential plants; this latter prediction is surrounded by considerable uncertainty.

A similarly uncertain but potentially more important effect is connected with some changes in average temperature. These might be caused by the fine dust stirred up by the explosions and by the absorption of various wavelengths of light by ozone molecules. Ozone absorbs sunlight, leading to an increase in temperature. A further increase is due to the circumstance that ozone also inhibits the reradiation by the earth in the longwave, invisible spectrum. The dust that is stirred up and the redistribution of ozone can give rise to consequences which are difficult to predict. It might create a temperature change—on the order of one degree on the average—that could last for a few years and could have further ecological effects.

It could be postulated that in a future war ten times the amount of nuclear explosives considered in the Academy report will be used. It is highly unlikely that any military operation would plan to use atomic weapons on this scale. If nothing else, the expenditures needed to prepare for such insanity are beyond the means of any country. However, even if such a further escalation could occur, it is still highly probable that the human race would survive.

One of my very imaginative friends, Herman Kahn (whose intellect matches his girth), wrote a book that had a great impact on people's thinking. Like most books, it was partially right and partially wrong.[11] At the moment, I want to point out where it was wrong. The title, *On Thermonuclear War*, is not as impressive as one of the topics of the book—"Doomsday Bomb." This topic gave the book considerable impact.

Atomic bombs were one thousand times bigger than the block-busters of World War II, and the hydrogen bomb surpassed the atomic bomb a thousand times again. When people see something

[11] Neither this book, nor this footnote claims to be an exception.

very big, there is a tendency to imagine that it is infinite. The consideration of the absolute, of the ultimate, even when there is no visible path to get to it, is seductive. It was natural to expect another bomb, bigger by another factor of one thousand. Such incredibly powerful bombs did not materialize—for a good reason. One can show that even bigger bombs would harmlessly disperse most of the additional power into space. When I confronted Herman Kahn with the question of whether he could actually foresee a "doomsday weapon" being invented, he answered that he couldn't. However, he added, no one could prove that it potentially couldn't exist.

Knowledge of recent developments in weaponry provides a sense of proportion that can be used to evaluate the possibility of a mega-mega-bomb. In the past ten to fifteen years, the actual change that has occurred has been the design of smaller, "cleaner" weapons. Such weapons have greater military effectiveness, that is, they would be useful in fighting against military troops. Another major change involves the development of counter-offensive weapons which have the potential for effective defense. The innovations that have occurred and that will be increasingly important are qualitative in nature, not quantitative.

If one compares the findings of the National Academy of Sciences' report with the widespread myth that a nuclear war would be the end of the human race, the discrepancy is glaringly obvious. The predicted physical changes would seem to be quite negligible compared to changes that would occur in the world of human hatreds and fears, in politics and institutions. In these regards, a nuclear war would have a great and dreadful effect on the lives of men and women—but they would stay alive.

In his comments about the Academy report, President Handler presents some complementary aspects. He asks about the questions which are not discussed. How does one know that widespread low-level radiation added to terrible "hot spots" in combatant countries might not weaken resistance to disease or lead to a spreading ecological change? It is known that atmospheric phenomena are full of trigger effects. How can we be certain that limited temperature changes may not result in disproportionately large climatic variations? Above all, how do we know that changes in ozone are the last surprises a nuclear war would bring? A few years ago, no one considered this change an important consequence of nuclear war.

The facts considered thus far suggest that the end of the human race is not to be expected—*on the basis of the knowledge humans currently possess.* If the apocalypse were to occur, it would be caused by factors that lie outside our present knowledge. A relevant question now occurs: What is really new about this danger to the human race? The gravest danger is the one that is not foreseen. In all the catastrophies of history, natural or man-made, the ultimate danger has always been in the unknown.

Non-Scientific Surprises: Less Pleasant

In the 14th century, approximately one of every three people inhabiting Europe and central Asia died of the black plague. Whole villages in England were deserted and were never again inhabited. The terror returned in three successive waves. Its means of transmission were unknown; it was almost without exception fatal if contracted. It would be unbelievable had the generations which endured this horror not been convinced that the end of the human race had come. The plague's virulence passed, and the human race survived—for unknown reasons. The Black Death was a natural catastrophe. It is hard for me to consider nature less dangerous than man, but since some do, I shall examine human cruelty.

Warfare has a long recorded history. One would like to hope that mankind is, perhaps, making progress toward finding other ways to solve differences. Perhaps we shall be endangered to a lesser degree by leaders able to convince or coerce people to violence. History does not yet suggest such a development, but it does offer evidence that the destruction of human life has always been limited by the attitude of people to stop killing, rather than by a physical inability to slaughter. The combatants, no matter how cruel or callous, have aims that are not best served by the general destruction and total extermination of the enemy. In a war fought by rifles, it is not the number of bullets that can be manufactured that determines how many people die. In most wars, the limit is provided by the lack of desire of the victor to inflict further damage.

According to tradition, there was once an exception to this rule. In 1219, the armies of Genghis Khan were in Persia, separated from their home base by tremendous distances. Perhaps because of the difficulty they would have had if forced to retreat, they did not want to spare *any* of their enemies; they tried to destroy all Persians. When they took a city, they not only killed everyone they could find, they left the city and returned a few days later to slay those who crawled out of the ruins. Perhaps there is no example of greater havoc in human history. It is deeply hoped that it will never be repeated. Yet, at least 10 percent of the Persian population survived.

There are differences between the past and future. By its very nature, the future is the only period that can really inspire fear.

Furthermore, in the past, the methods of killing have been more familiar. The assertion that all people would be killed could be at least partially answered by arguments that most people could understand. Today's methods of destruction are unfamiliar to the vast majority of people. They do not even know how to ask questions about such destruction nor how to doubt any asserted exaggeration. The idea of the apocalypse is not new. The Four Horsemen have long been identified. There is now a Fifth Horseman who is more terrible because he is unknown. It is one of the obvious duties of the scientific community to describe him as accurately as possible. Only in this way will he cease to be unthinkable and will he become avoidable. Until this happens, the new mythology that the days of mankind are numbered will haunt people's minds and will confuse their reasoning.

Assuming this discussion has enabled some readers to put aside thoughts about the end of the human race, it is still suggested that mankind may survive but will be thrown back to the Stone Age. Significant counter-examples exist. After the Second World War, the material destruction in Germany and Japan was very extensive. It was not 100 percent, but it approached that level. In addition, both countries lost a considerable number of their people. Both countries recovered quickly, in a manner that has been called an economic miracle. (In this context, a miracle is something that the present state of economic science is not apt to predict.) In a modern state, the value of all goods that can be bought and sold is approximately equal to the value of three years' production. This fact is ample evidence that a truly destructive war would not be the economic end of civilization. If knowledge of science and technology were lost, the recovery would then involve a longer period.

Having considered wars and weapons, I now want to consider purposes. I have lived through two world wars. Their contemporary impact and their long-range consequences are vivid in my mind. I can see nothing more important than to avert a third world war. Not only do wars create incredible suffering, but they engender deep hatreds that can last for generations. The terrible consequences of the loss of freedom are much less appreciated since they are not visible and have not been experienced by most people living in democratic nations.

Among the many things that science and technology can accomplish is human control not only over nature but over other humans. This control can be abused, or it can be used beneficially. A major problem facing democratic nations is the development of rules under which the possibilities of technology shall not be misused. These rules would serve to place valid, consistent and understandable limits on the applications of science. Computing machines have been used as excellent laborsaving devices in many walks of life. In a totalitarian state, they can be used to strengthen the bureaucracy and make control over people practically complete. They can constitute one effective element in depriving people of privacy. It is necessary—and it is possible—to increase the safeguards of freedom when technology makes it easier to suppress freedom. In democratic countries, there are a great many examples of how freedom can be safeguarded.

In the international arena, it is much more difficult to avoid the misuse of technology. The most simple procedure would seem to be to establish rules that will abolish destructive weapons. Rules, however, in order to be effective, need to be enforceable. From this point of view, the surest way to maintain peace, stability and decent conduct is to have power in the hands of those who stand for decency, stability and peace. (It is also advisable to keep the power in the hands of many, for it is more difficult for large numbers of people to agree on gross abuses.)

This is a simple enough prescription provided one can agree on who it is that wants the right things. In one of Aesop's fables, the wolf has to devour the lamb in self-defense because the lamb is an aggressor: he disturbed the water in the stream from which the wolf drank. (As it happened, the lamb was downstream from the wolf, but such details are often considered extraneous.)[12] On a similar basis, one can make a case for a country needing to launch two world wars because of a vulnerable geopolitical position. Those who know Western Europe will realize that no part of this remarkable portion of the earth wants a third repetition.

Applying A Complementary View

There are a multitude of international problems which confront us today. In seeking solutions, I think it helpful to use a method drawn from the pursuit of simplicity. Each of our problems should be considered from two opposite, one could say, from two complementary points of view. There is little controversy about the idea that the world is interrelated, that there are problems which have mutual effects and require global solutions. There is controversy about which problems fall into the international-concern category. I would recommend that the problems be sorted and solutions for them sought on

[12] Aesop himself was hurled to death from a rock near Delphi because he was considered a disturber of the peace.

the basis of a complementary approach. This recommendation is so simple that it appears almost empty. I am suggesting that Niels Bohr's approach should be widely disseminated and within its inherent ample limits, compromises be found between the opposing facts and situations found in the real world.

For example, self-determination as a right of individual nations must survive. Yet cooperation must be established if the welfare of any of the world's people is to be maintained. Which should take precedence can only be decided case by case, and it must be subject to review. Sometimes one will dominate, and sometimes the other. The two extremes—complete self-determination and total coordination of nations—must be held in a complementary fashion in order to arrive at the best, possibly the only practical, course of development. However, this flexible manner of thinking is not easy to achieve. Traditionally, people adopt one point of view and stick to it until events prove the view untenable. At this point we customarily abandon the original position and adopt the opposite, equally faulty approach.

I am proposing that we learn to consider issues from two apparently contrary views at the same time and then choose the mixture that best suits the individual situation. No doubt this idea will be subject to charges of double standards and batting for one side while pitching for the other, of inconsistency and possibly even worse. Yet, just as it was necessary to adopt complementarity in atomic science to obtain understanding and simplicity, it seems equally imperative to adopt such an approach to global problems.

The third edition of Webster's Dictionary offers an example of what I am proposing. This edition reflects a substantial change of philosophy from the early concept of a dictionary as a "standard." The third edition is offered as a compendium of the common usage of words. The shift can be viewed as one from authoritarianism to a new freedom and democratization in the use of words. Obviously, such a shift could be carried too far, and it has led to some ludicrous situations.[13] However, this approach is a better reflection of reality and thus serves the purpose of a dictionary more widely and adequately.

Before going further in a discussion of problem solving, it seems necessary to acknowledge an important limitation. The longing for perfection, the striving for the ultimate truth which seems inherent in our thinking introduces the dangerous idea of a final and complete

[13] Rather than define *bi-monthly* from the standpoint of setting a standard, the third edition says that the most common usage is "once every two months" and the next most common meaning is "twice a month."

solution to problems. Just as I have come to understand a limitation in our measurements as inherent in the reality of the atomic world, and as I accept a limitation of the value system in politics, so I conceive of successful solutions as having a temporary nature. As W. H. Auden expressed it,

If thou must choose
Between the chances, choose the odd;
Read the New Yorker; trust in God;
And take short views.[14]

It would be wonderful to be able to solve problems once and for all. It is very doubtful that this is possible. It has often been claimed, has never been accomplished, and historically it has proved to be a destructive idea.

Using the method of complementarity, I shall look at the attainment of peace. One can take the position that since wars are fought with weapons, societies should do away with weapons. If there are arms, they may acquire a dynamism of their own and bring destructive conflict to those who possess them. "He who lives by the sword shall die by the sword." I have heard a distinguished anthropologist assert that weapons are addictive.

A different approach to the question of how to preserve peace would suggest that the emphasis should not be in abolishing or limiting the weapons but on ensuring peace by every means at one's disposal. Consider the situation in 1945. The United States could have established absolute control of Germany and Japan as they were conquered enemy nations. We chose a different course. Our former enemies were revived as friendly democracies. This resolution of conflict has few, if any, precedents in history. It is now possible to argue that love of peace and effective pursuit of it are more than pure imagination. Consistent and well-intentioned foreign policy, co-operation and openness can help us preserve peace.

A more sharply different approach is also important to acknowledge: that power in the hands of those who prefer peace is vital to its preservation.

Human rights, no less than peace, must be examined from complementary positions. It is a great and noble temptation to support human rights as an absolute unquestionable principle upon which to base one's actions. I personally value human rights most highly. Unfortunately, even this great good cannot serve as *the*

[14] From *Under Which Lyre* by W. H. Auden, copyright June, 1947, by *Harper's Magazine.* Emphasis added.

fundamental basis. There are unavoidable conflicts of interpretation and of feasibility. Shall we hide from ourselves the horrors of concentration camps in those countries where we can not hope to abolish them? In politics, moral values and practicality cannot be completely reconciled. This does not mean that the value should be discarded. But it does mean that there can be no exceptions to the application of complementarity. Without exception, the need for exception is universal.

Furthermore, freedom, the ultimate expression of human rights, cannot be considered a gift if it is to be preserved. When freedom is a gift, it is not appreciated. But when people acquire their freedom, either by struggle or by work, they can form a nucleus to help a wider group attain this privilege, a privilege which carries responsibilities and demands sacrifices. However, it is first necessary to bring about a close cooperation among those nations that have a common heritage in freedom. Fragile freedoms are better protected when all free countries work closely together. To the extent that people hope to form such a nucleus, they must be aware of the fact that freedom is imperfectly secure, that the alliance for its defense is imperfect, and the future of the alliance is far from guaranteed. Yet it is often forgotten that inconsistent behavior, such as the sudden abrogation of treaties, may have the consequence of rendering any real collective endeavor in the future impossible. All these self-evident statements are old. Technology has given them relevance on a world-wide scale. Thus it becomes even more difficult to implement them.

What follows from these thoughts is the need for a consistent program using the rapidly increasing power of science and technology to acquire a better and more secure future; among many other things, this would include defense. I would not argue for the defense of just the United States, but rather for the defense of freedom in those parts of the world where freedom is firmly established. The extension of freedom, though certainly to be desired, is much more subject to doubt than the preservation of freedom. What is more, the proper moment to consider the problem of defense is now. One must bear in mind the fact that many people who have at one time enjoyed freedom are no longer able to do so. When one has suffered extensive defeats, it is not wise to spread one's remaining forces too widely.

Modern methods of defense preparedness are inevitably based on science and technology. These new developments are usually kept secret in order to "protect" them from their use by enemies. However, secrecy hampers cooperation in the free world. There is no greater strength than unified cooperation. Protective secrecy is actually the opposite—it damages the ability of free people to cooperate and survive. By using the opposite of openness, we reap the results: a weakening of freedom. Therefore, I am arguing for the abolition of most of our national secrecy.

Here again to get the correct perspective, one must consider an opposite argument. If democracies publish everything they know, the totalitarian regimes will know all that plus everything else they have learned through their own secret efforts. Therefore, they will clearly be ahead in science and technology.

The issue of secrecy is not that simple. Most secrecy tries to protect an idea. Such secrecy may succeed for a while in a totalitarian nation, but it never succeeds in a free country. Protecting scientific ideas is particularly pointless because the execution of current technology requires so much more than ideas: it requires extensive and detailed applications. These applications are the real secrets, and they should be communicated only to one's allies. These practical details actually cannot be given away easily except by working on the problem in a joint fashion. (Students of quantum mechanics are known to believe that this subject is one of the better kept secrets.)[15]

Expanded freedom in the exchange of ideas would greatly increase technological progress in the free countries. It would strengthen their alliances and the well-being of all peoples by speeding the peaceful use of everything learned through weapons research. The advantages would far outweigh the fact that totalitarian states would know more. They still wouldn't know the details which make execution possible.

What is perhaps equally important is that openness will have a great effect in the nations where freedom is denied. There are in these countries many people who are struggling for a more open system. It is easy to think of many highly visible examples. Increased openness would aid the determination of people who continue to think in those parts of the world where speech and thought are shackled.

A totalitarian state can be effective in keeping secrets, at least for some time. However, the existence of secrecy itself cannot be concealed. The fact that a government is not open will quickly become obvious. Aid to any country which did not adopt a similar policy of openness in regard to science could be denied. If cooperation through openness begins to give visible results, the inducement for countries to drop secrecy may increase.

[15] The reader who has attempted the appendices to the third chapter no doubt sympathizes with this opinion.

A seemingly endless number of international problems remain. The first-mentioned problem of the independence of the nation-state is of prime importance. But the world has become so closely interconnected, the fate of nations so intimately interwoven, that the idea of sovereignty—each nation determining its own course—can obviously only hold to a limited extent. Many of the infinite number and variety of problems, for example, disease, freedom of speech, international commerce and population growth, concern all of us no matter where on earth they occur. An impressive amount of progress has been made on disease control. In this, even an imperfect organization like the United Nations has made real progress.

Freedom of communication, the opportunity to consider ideas, as opposed to censorship and secrecy, is vitally important to humankind. Censorship, like identification of an aggressor, can take many forms. The *Reader's Digest* is translated into most languages and read widely around the world. It is remarkable that some consider this publication a form of imperialism and believe that it should therefore not be circulated. Regardless of whether or not one is an avid fan of a particular magazine, one should certainly argue for the right of people everywhere to read it if they wish. This example has received little attention, but it should be taken seriously.

The case of interational commerce is much more difficult, and I will acknowledge this by not discussing it. Not many years ago, economists said that the laws of economics make a cartel of raw material producers impossible. Since then OPEC, in apparent ignorance of this particular law, went ahead and disproved it. Because my knowledge of economics is vastly inferior to that of the economists who made the prediction, I shall now turn to an area of social science where at least some obvious mathematics is involved.

Population Growth

Another problem for which global cooperation is advocated is that of the world's accelerating population growth. Medical advances, basically a branch of technology, have greatly diminished infant mortality, even in countries where other technology is largely absent. This, combined with the overall control of infectious diseases, explains much of the rapid population increase the world is experiencing today. One cannot be opposed to the cause; yet, one is worried about the effect. There are now concrete reasons to fear the consequences of overpopulation, and the most immediate one is starvation.

Ever since Thomas Malthus described a balance between food increase and population in the 18th century, his point of view has been commonly accepted. Malthus said,

> Population, when unchecked, increases in a geometric ratio. Subsistence increases only in an arithmetical ratio. A slight acquaintance with numbers will show the immensity of the first in comparison with the second.

This is double nonsense. In the 180 years intervening, world population has increased at a rate greater than geometric, and the means of subsistence have increased vastly beyond an arithmetical progression.

Human inventiveness in the fields of science and technology have outstripped human fertility. When Malthus said that only war or starvation could maintain the balance between population and food supply, he failed to allow for new ideas, human desires and, most of all, for technological developments. For this he may be forgiven; at the beginning of the Industrial Revolution, it was difficult to anticipate its achievements. It is now very clear that science and technology have greatly contributed to human survival.

Malthus' second nonsense is harder to forgive. There is no evidence in history, not at his time or after, to justify the use of the mathematical expressions he cited. "A slight acquaintance with numbers" aptly describes Malthus' proficiency. The facts not only differ from Malthus' projections, they are amusing in their wide variance.

The idea of a geometric increase in population is based on the assumption that rate of increase in population should be proportional to the number of people present. If one looks at the history of the past 400 years, it is immediately obvious that this pattern is not valid. The doubling time of world population has changed quite rapidly. Beginning in 1650, doubling occurred successively in 200 years, 80 years, and 45 years. A geometric increase predicts a constant doubling time.

If Malthus had projected backward, using the doubling time available to him of 200 years, he would have calculated that there were only 50,000 people in the world in the days of King Solomon (about 1000 B.C.) and that at the time when the Great Pyramid of Khufu (Cheops) was built in 2560 B.C., there would have been only 170 people in existence. It is very hard to believe that Malthus ever attempted a serious numerical proof of his theory.

There is a formula proposed by zoologists which agrees with the observed data to a reasonable extent and which seems to hold for most animal species if they are isolated and have ample sustenance. If N(t) is the number of people alive, and t is the time in years since the birth of Christ,[16] the equation

$$N(t) = \frac{200 \text{ billion}}{2030 - t}$$

gives quite good values for the last 400 years and reasonable values for earlier times for which reliable data are lacking. In Figure 5.1., estimated populations are compared with the proposed formula. Predictions derived using Malthus' theory are entered in the last column of the table.[17]

Figure 5.1.

POPULATION TABLE

DATE	*Historical Estimates*	*Proposed Formula*	*Malthus' Theory*
B.C.			
9000	a few million	18 million	less than one person
2650	(?)	43 million	170
1000	(?)	66 million	50,000
A.D.			
Christ	200 million	99 million	1.65 million
1000	275 million	194 million	53 million
1650	530 million	526 million	503 million
1750	720 million	714 million	711 million
1800	912 million	870 million	870 million
1850	1,131 million	1,100 million	1,005 million
1900	1,630 million	1,534 million	1,196 million
1920	1,811 million	1,818 million	1,281 million
1930	2,015 million	2,000 million	1,327 million
1940	2,249 million	2,222 million	1,373 million
1950	2,509 million	2,500 million	1,422 million
1960	3,027 million	2,857 million	1,472 million
1970	3,678 million	3,333 million	1,523 million
1980	4,330 million	4,000 million	1,577 million
1990	5,188 million	5,000 million	1,633 million
2000	6,130 million	6,667 million	1,691 million

[16] For dates before Christ, t has a negative value.

[17] Malthus might say that wars and famines account for the lack of agreement. That may hold for the period before 1650, but it makes the application of his statement meaningless.

In addition, the proposed formula has an interpretation which is more rational than the spontaneous self-reproduction of Malthus' model, although it is hardly compatible with human reproduction scenarios. The above formula can be obtained by assuming that the rate of increase is proportional to the square of the number present rather than to the number of people itself (as Malthus implied). This hypothesis is not completely absurd; it is proportional to the number of men times the number of women. One can with some justification consider procreation as primarily due to "binary collisions."

Obviously the formula will break down before the year 2030, for the formula predicts that there will be an infinite number of people in the world in 2030. But up to the present, it gives a more adequate basis for numerically estimating population than Malthus' theory. It is not understood what really limits population growth. But by studying deviations from the formula, one might find a way toward practical understanding.

It is fortunate that local self-limiting processes occur in connection with population growth, because current evidence makes it appear hopeless for an international organization to attempt to make rules to limit growth rates. No country will permit interference in such matters as long as it has a shred of sovereignty left. It is equally apparent that the very countries which protest when given advice on population matters accept the advice tacitly and themselves attempt appropriate actions.

Technology As Part of the Solution

None of the factors I have discussed thus far—defense preparedness, cooperation and openness, taking a complementary view—is in itself sufficient to ensure that human life will be dignified throughout the world. Today only one out of every four people has a tolerable standard of living. More than three billion people are living in degrading poverty! They have no use for freedom for they have no control over their lives. Technology can change this situation, but only if it is pursued with vigor.

I am appalled at the lack of perspective demonstrated by some social scientists. In *The Affluent Society*, Kenneth Galbraith proposes that all the problems of poverty are solved except the problem of distribution. Mr. Galbraith seems to forget the billions of people living in the world under conditions of great, indeed, scandalous poverty. In his book, a total of two pages is devoted to problems that extend beyond the boundaries of the United States. One would not have expected such a determined isolationist point of view from an economist.

Since human beings appear to need a purpose in order to lead rewarding lives, it may be that nations also need purposes. I propose for us, as members of an affluent nation, that purpose should be to help in the development of the world as a whole. In principle, we already espouse this concept; in practice, we have done far too little. And the prospect is that our prosperity will be dissipated by the neglect of technology without having given enough help to others.

There are some honorable exceptions to this history of neglecting both technology and our neighbors. In connection with the basic issue of food, the Rockefeller Foundation established the Rice Institute in Manila, a major step towards the prevention of starvation. The new grain development research in Mexico City, the ocean resources development in Hawaii, and other similar activities around the world continue to pursue this admirable purpose. Today, 7 percent of all food comes from the waters of the earth, but this 7 percent represents 20 percent of the protein which people consume. Protein shortage is the particularly critical issue. More comprehensive utilization of the oceans should be considered.

If protein collection from the ocean is restricted to fishing, it is not easy to exploit the oceans efficiently. Indeed, using only this method, mankind would quickly approach the limits of the resource. People are still in the Stone Age when it comes to fishing techniques. We harvest at random without trying to sow. What is needed is a method of cultivating ocean protein so that mankind can move from the hunting-gathering stage to "mariculture" (which would correspond to agriculture in the sea). If we hope to increase the available protein resources so that every human has a full potential for development, we must repeat the progress of our early forebears and exercise the complexity of preliminary effort and replenish the fish population before reaping a reward.

A great deal is known about freshwater fish, including their diseases and the best way to grow them in fishponds. This resource is limited in comparison to saltwater fish, however, and knowledge about fish of the oceans is primitive. The cultivation of salmon—growing the young under crowded conditions and releasing them into the ocean able to survive—is an important current practice. But it is an exception, unfortunately, rather than the rule. The sedentary varieties of shellfish can be cultivated in highly effective ways. This practice is underway today in Japan. But the application of similar technology to the many edible varieties of free-swimming fish that occur in the ocean is an unsolved problem. The current knowledge of saltwater fish disease can be compared to the knowledge of human disease before Pasteur.

Of course, this is only the beginning; the actual practices of ecology will rely on answers to many questions. Can one use atolls in the ocean (which can be protected against predators) as huge fish hatcheries? Should one consider the inedible predators among fish as a sort of weed to be rooted out? On the other hand, dare one upset nature's balance by introducing changes in the population of the ocean?

This list of questions will surely be answered in the negative by most environmentalists, and, indeed, they have been strongly opposed to the decimation of whales, as one example. Certainly, it is necessary to look at the problem from both points of view. But considering the unavoidable increase in the human population until at least the year 2000, the problem of avoiding hunger and malnutrition must be given the greatest weight. It is particularly important to note that the hydrocarbons of the ocean are indigestible by humans. The hydrocarbons must continue to come from land; on the other hand, the oceans could supply the all-important proteins.

The technical problems of mariculture might not be easy, but the political problems will surely be more difficult. The reason for this is clear. If one nation sows in the ocean, who is then allowed to harvest? The questions regarding a "Law of the Sea" are both urgent and difficult. So far the primary participants in groups considering the questions have been politicians. The discussions have remained barren. One of the ingredients of a solution to these problems is to include people with technical expertise who are well practiced in using the techniques of scientific pursuit. A hopeful exception to the pattern of exclusively non-technical participation in political affairs can be currently observed in the State of Washington where Dixy Lee Ray, a marine biologist, is governor. The manner in which this state approaches and solves its problems during the next few years may demonstrate the validity of this suggestion.

There is a parallel to the problems of the oceans in the problem of the atmosphere. On the whole, we know the basic laws of this part of our environment, but we do not understand the complex ways in which they operate to give rise to instabilities. Small initial perturbations can lead to big consequences, and this circumstance makes weather prediction a lot more difficult. Major progress is probably close at hand if atmospheric research is emphasized. Humans have attained the wide horizon of the angels through satellites, and computers allow reasonably accurate weather predictions to be made as long as they are given the present global state of the atmosphere. Eventually, this combination should allow the weather patterns that are the "triggers"—that influence the changes in wind and rain and sunshine—to be identified. Droughts and, perhaps, even floods could become avoidable.

Full Circle: Another World's End

In an earlier section I discussed the most prevalent "world's end" myth. Today people seem to be well supplied with such theories. My comments on the weather remind me of another instance. In a recent book, *The Hunting Hypothesis,* the late Robert Ardrey presented his particular version of gloom. It is the coming ice age. Indeed, Ardrey had good reason to fear an ice age even more than he feared human folly and the cruelty of aggression.

We have an excellent scientific record of the glacial periods for the last 700,000 years. In each case, the times during which the glaciers relaxed their grip over the earth's surface seem not to have lasted longer than about 10,000 years. Recent human history has had its 10,000 year share of sunshine and a little more. There is some evidence indicating that the next ice age might overwhelm us within a single century. Ardrey saw no escape except for a small hardy band of people who will have to endure the worst.

Ardrey has thought and written extensively, wisely and eloquently about the past. But one needs to think more about the future, and one needs to include mankind's greatest servant, technology, which was never at human command in earlier ice ages.[18] No one knows what causes an ice age, but it is known that during such a period the earth either receives or retains a few percent less energy from the sun. That we are always so close to disaster is a frightening thought.

The use of complementarity allows one to look at the same fact in a different way. If the world is always so close to disaster, the balance between well-being and destruction could be tipped with some ease. Technology has already performed many miracles. Could it turn back a glacier? I am convinced that it could.[19] Of course, I don't know how this could be done, but, in spite of my ignorance, I shall offer a couple of examples of possible solutions.

Humans have managed to hurl fairly massive bodies into space. To increase the world's energy supply, mirrors constructed of extremely thin materials could be orbited around the earth. They could be kept properly oriented by spinning them around a slowly changing axis so

[18] I had the great pleasure to correspond with Ardrey about these questions just a few days before his untimely death.

[19] Such a statement will undoubtedly label me as an optimist. Therefore, I shall repeat some definitions: a pessimist is one who is always right but gets no enjoyment out of it; an optimist is one who imagines that the future is uncertain. I think it is a duty to be an optimist for one is apt to act if the future is seen as uncertain.

that they reflect sunlight to the right spots on earth. Which are the right spots? Every winter a heavy pool of frigid air forms over the pole. Winds blow around this pool, raising waves just as a wind raises waves on water. These waves of air are exceedingly long; the whole circuit around the pole is composed of only three or four wavelengths. These are called planetary waves (they encircle the planet) and are actually quite unstable. They grow, energized by the wind. In the end, one crest with maximum amplitude breaks off and a big chunk of arctic air drifts into the temperate zone. The mixing of this cold air mass with warm moist air gives rise to the all-important winter rains.

If people could influence the instability of the waves on a huge scale, they would have made a long step toward weather control. One may direct the sunlight from the mirrors toward the appropriate spots along the interface between the cold pool and the westerly winds (this interface is called a cold front or arctic front), thus effectively warming a critical spot where instability may originate. It is quite possible that, by these means, control of an ice age could be established.

I would guess that such a project might cost as much as $1,000,000,000,000. Is this sum too great to be borne? It is much less than the gross national product of the United States for a single year. If the issue were the salvation of human civilization, then a price even ten times greater would not be excessive.

There is a less expensive solution which, indeed, is already in operation. Increasing the amount of carbon dioxide in the atmosphere by burning more fossil fuel also increases the average temperature. This increase has been happening over the past centuries. The carbon dioxide acts like a blanket that keeps the earth warm. It is estimated (taking into account that some carbon dioxide is dissolved by the surface layer of the ocean and eventually precipitated as calcium carbonate to the ocean floor) that, by the year 2050, the carbon dioxide in the atmosphere will have doubled. The average temperature might rise anywhere from two to seven degrees Fahrenheit. The glacier-fighting might even be overdone. Furthermore, all these results are as yet uncertain.

But have I really answered the question about survival in this case? People have reason to be concerned about the conflict that could be engendered if a nation tried to influence the seas, let alone the climate. I do not really mean that the Sierra Club could successfully defend the slogan "Save Our Glaciers," but rather that no change imaginable will benefit everyone, let alone benefit everyone equally, and this is, indeed, cause for concern. If this is an occasion for strife then there will be strife. So, in the end, one comes back to the question of whether technology can be used in a reasonable manner.

When people are confronted with an ultimate and actual catastrophe, reason may prevail. But complementary facts cannot be evaded. Technology without reason will probably be worse than the total absence of technology. Reason without technology finds us helpless before the powers of nature. Reason creates technology, but reason is not always present in its employment. Mankind can look forward to the future with hope only if all the human gifts that distinguish us from other living creatures are used. It is even possible that the menace of oncoming glaciation will provide human beings with sufficient motivation to unite against a common menace of nature.

And A Beginning

The preservation of peace and the improvement of the lot of all people require us to have faith in the rationality of humans. If we have this faith and if we pursue understanding, we have not the promise but at least the possibility of success. We should not be misled by promises. Humanity in all its history has repeatedly escaped disaster by a hair's breadth. Total security has never been available to anyone. To expect it is unrealistic; to imagine that it can exist is to invite disaster.

What we do have in our technological capacities is an opportunity to use our inventiveness, our creativity, our wisdom and our understanding of our fellow beings to create a future world that is a little better than the one in which we live today. This should be our present aim—to reconcile science and survival. I want to emphasize that survival is not simply the fact that some people will continue to live on the earth. That much I consider a certainty for the foreseeable future. It seems better to aim not only for the survival of people but also for the survival of human dignity.

In the attempt to survive, in the attempt to reconcile what seems contradictory, people may succeed only if they manage to find the proper amount of simplicity in a seemingly complex world. One needs truths that are accepted as evident. I would offer these:

> That no limits are set to human knowledge;
> That knowledge leads to power;
> That power can be used for the benefit of everyone.
> If we believe more than this, we may be fools;
> If we believe less, we are cowards.

No endeavor that is worthwhile is simple in prospect; if it is right, it will be simple in retrospect. The pursuit of simplicity in science leads to understanding and beauty. In human affairs, it may fulfill our most desperate need: the survival of a civilized human society.

NOTE TO THE READER

A book about science is of necessity incomplete. Some readers may regret this. As a partial remedy, I was tempted to introduce appendices. However, the problem with appendices is that they are unnecessary for the expert and uninteresting for the majority. Therefore, I arrived at a compromise: invisible appendices.

My plan is this. The appendix will be cited in the usual way in the text. The appendix will name the subject to be explained, and a way of writing the appendix will be outlined with utmost brevity. Those very few people who start as laymen and are stimulated to become experts may then attempt to write their own appendices.

If anyone succeeds, s/he is invited to send the completed work to the author, in care of the publishers,

Pepperdine University Press
Malibu, California 90265

and the author will in his turn consider what the reader has written. The reader is encouraged to hope that the author will reply.

This procedure may accomplish two purposes: to save paper and to stimulate thought.

Appendices

APPENDIX I: MEASURING THE DISTANCE OF THE SUN FROM EARTH

The earth-sun distance is very great compared to the earth-moon distance. At the time the moon is in the first or third quarter, the sun-moon-earth angle is a right angle. The triangle formed by the sun, moon and earth is quite elongated. A small misjudgment of the precise time at which the triangle becomes a right triangle results in a big mistake when calculating the ratio of sun-earth to sun-moon distances. In this appendix, the figure of the earth, sun and moon in its first (or third) quarter should be drawn. The quantitative relationships between the angles and the ratios of the distances and errors introduced in the distances (due to errors in the measured angles) should be discussed.

APPENDIX II: DERIVING KEPLER'S SECOND LAW AND THE CONSERVATION OF ANGULAR MOMENTUM

Kepler's Second Law (equal areas are swept in equal times) can be derived as follows. Take a short time interval. Assume that there is NO force acting on the planet. Show that in two equally short times, equal areas are swept out. Then show that if the force is directed towards the sun, the result remains unchanged (which proves Kepler's Second Law). Show further that this Second Law is equivalent to the conservation of angular momentum. Remember that momentum is mass times velocity, and the angular momentum around a location (in this case, the sun) is the distance from the location multiplied with the component of the momentum perpendicular to the line connecting the moving body with the sun. (For the expert: angular momentum, like momentum, has direction which is defined to be perpendicular to the momentum and connecting line.)

APPENDIX III: DERIVING THE INVERSE SQUARE LAW FROM KEPLER'S THIRD LAW

Assume (in a slightly erroneous way) that the planets move in circles around the sun. Assume that the acceleration (that is, the change of velocity in unit time), due to changes in the direction of the motion, is proportional to the gravitational force between the earth and the sun. Find the relationship between the force, the velocity and the curvature of the orbit which is related to the radius, i.e., the distance from the sun. Use Kepler's Third Law according to which a^3/t^2 is a constant, and then show that the force turns out to be inversely proportional to the square of the distance.

APPENDIX IV: DERIVATION OF KEPLER'S FIRST LAW

The proof of Kepler's First Law can be given in five steps.

1.) Prove the conservation of energy, that is, prove that if you take the change of potential energy (or force times the distance traveled in a direction parallel to the force) and add to it the change in kinetic energy (1/2 the mass times the velocity squared), this quantity remains constant.

2.) Use the conservation of angular momentum.

3.) With the help of these two conservation laws, construct an equation relating the component of planetary velocity parallel to the line connecting the planet with the sun, the planet's distance from the sun, and its total energy. The result is called a differential equation.

4.) Solve this equation which will give the way in which the distance from the sun varies with time.

5.) Figure out, using the conservation of angular momentum, how the angle changes with time. Show that the resulting curve is the one given in Kepler's First Law.

Remember that Newton took two weeks.

APPENDIX V: DETERMINING THE COORDINATES OF THE POINT OF ORIGIN

It is suggested that the reader figure out the coordinate values of the point of origin, that is, the intersection of the **X** and **Y** axes. If s/he cannot do it, s/he should perhaps skip this chapter. (Hint: **x** = ?, **y** = ?)

APPENDIX VI: CONSIDERING THE ARRIVAL OF LIGHT FROM DOUBLE STARS

Using the example of the double star at a distance of 100 light-years from earth, assume that the two stars (which form the double star) and the solar system lie in the same plane. Further, assume that the two stars are moving on strictly circular orbits with one ten-thousandth of light velocity, have equal mass and circle each other with a one year period. Disregard the constancy of the speed of light. Then according to simple addition of the velocities, the light from the approaching star will have a velocity of 1.001 **c**, while the light from the receding star will have the velocity of .999 **c**. In 100 years, the light from the approaching star will arrive sooner than from the approaching star. The reader should figure out how the signals from the two stars would be distorted upon arrival.

Assume the same conditions for the double star, but assume a greater distance. How great must the distance be so that the light from the star approaching at maximum velocity arrives at the same time as the light from the same star arrives half a period later when it recedes from earth at maximum velocity? Figure out the resulting tangle in the observed signals. Figure out under what conditions a similar tangle would result if the stars were 200 light years distant.

APPENDIX VII: SPACE CURVATURE AND THE INVERSE SQUARE LAW

Applying the idea discussed in connection with the two-dimensional curved surface of the earth on page 71 to four-dimensional time-space, and using Einstein's law of gravitational red shift, one can derive Newton's $1/r^2$ law about which Newton refused to hypothesize.

The purpose of this Appendix is to give the reader an impression of the connection between four-dimensional curved space and Newton's inverse square law. The steps indicated cannot be carried out except by an expert or a genius.

1.) Consider a short arrow which is called a space interval, **dx**.

2.) Carry this interval around a closed quadrangle. To do this one should first wait a short interval of time (Δt), which, however, is long enough so that $c\Delta t$ is great compared to **dx**; second, make a displacement (Δx) which is of the same order as $c\Delta t$; third, turn back the clock, that is, undo the original waiting period (Δt); fourth, move back to the original positions, that is undo the original displacement (Δ**x**). There is curvature if, as a result of these operations, the small space interval **dx** has turned into the same interval plus an exceedingly small time interval. The whole process can be performed in an equivalent but easier way if instead of completing the loop one moves the interval from the original point to the diagonal in two different ways: first wait, then move, and alternatively first move, then wait. Finally, compare the two resulting intervals.

3.) Note that there will be a difference between the two last-named routes in the presence of a gravitational field. Depending on which of the two routes one considers, one has waited at different places. Due to the red-shift, the waiting periods, if they are the same in the two different locations, will give a difference in the final time interval adding an exceedingly small time difference to the original **dx**.

4.) Perform this operation with the same interval **dx**, but use three mutually perpendicular displacements, Δ**x**, Δy and Δ**z**. Then add the results.

5.) Einstein assumes that the result is proportional to the density of matter in the region where the operations were performed.

6.) Show that if the result is zero (in empty space), this does not mean the absence of curvature (which was obtained before adding the results of the three mutually perpendicular displacements). It only means that the lines of force which represent gravity cannot start or end in empty space. The inverse square law follows from this picture of the lines of force.

Note that both Newton and Einstein made an assumption. Newton assumed the inverse square law. Einstein assumed a relation between matter and curvature. To the unmathematical mind, Newton's assumption will appear more simple. But what is simple is not so easy to decide. One fact is certain. Einstein, by making his assumption, explained many actual phenomena of which Newton never dreamed. Einstein greatly refined Newton's theory which now appears as an approximation valid for weak gravitational fields.

APPENDIX VIII: REFRACTION OF LIGHT IN WAVE AND PARTICLE THEORY

Newton assumed that the energy of the light particle did not change as light entered the water, but the kinetic energy and the total momentum did. At the same time, the ratio of momentum values in air and water remains the same regardless of the angle of incidence.

It is instructive to consider his derivation of the law of refraction which announced that for various angles of light incidence, sin i/sin r is constant. Here i and r are the angles that the incident and refracted light include with the perpendicular to the water-air interface. (See Figure A.1.) The expression *sine* is defined with the help of a right triangle. (See Figure A.2.)
The proof proceeds in the following steps:

1.) Assume that the ratio of the momentum values in air and water is independent of the angle of incidence;
2.) Note that only the component perpendicular to the water-air interface can change since no force acts parallel to the interface.
3.) Write sin i and sin r as ratios between momentum components parallel to the interface to total momentum values.

In order to repeat the proof according to Huyghens, assume the proportionality of wave number and momentum, and also the proportionality of each wave number component with the corresponding momentum component. (A wave number component is the number of waves one encounters in unit length if one proceeds in the direction of that component.) Assume further that the ratio of total wave numbers in air and water (along the direction of propagation) is the same, independently of the angle of incidence. (This corresponds to the fixed change of momentum in the particle theory.) The wave number along the water-air interface cannot change. (Otherwise the wave fronts would become discontinuous.) Then the proof proceeds in the wave-theory just as it did in the particle theory, repeating the steps above.

Notice also that according to Einstein's theory of relativity, momentum and energy are connected in the same manner as wave number and frequency.

APPENDIX IX: REFLECTION OF LIGHT FROM A MOVING MIRROR IN WAVE AND PARTICLE THEORY

Consider the particle theory of light, and calculate the energy change due to reflection by a moving wall.

1.) Calculate the momentum change of the wall as the sum of the momentum given to the wall by the approaching particle and the momentum due to the recoil from the reflected particle.
2.) Calculate the energy loss of the wall caused by the momentum imparted to it in a direction opposite to its motion.
3.) Using conservation of energy, recognize that the particle must have gained the energy which the wall (or mirror) has lost.

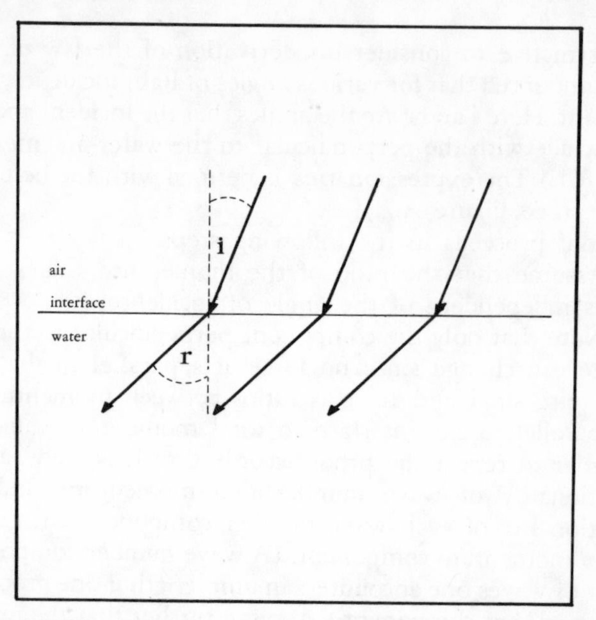

Figure A.1. The Refraction of Light Passing From Air Into Water.

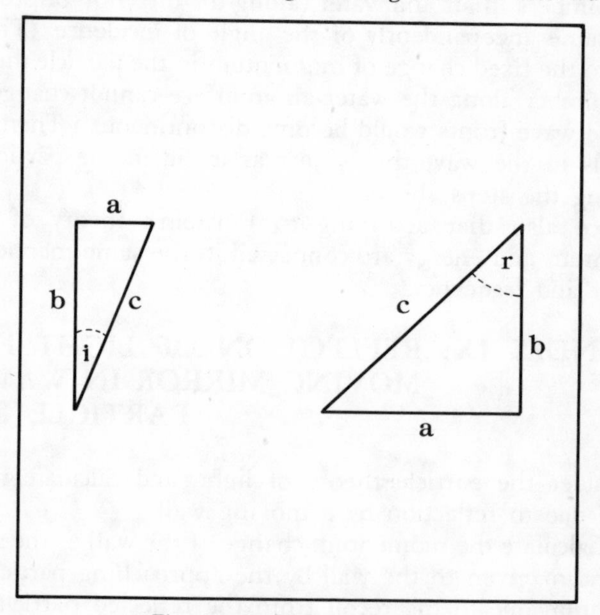

Figure A.2. Defining a Sine.

The sine is the line opposite the angle which is named (sin i) divided by the hypotenuse: a/c.

Now turn to the wave picture, and consider that all wave crests that have arrived in the incident wave must leave in the reflected wave. At first you may think that the two waves must, therefore, have the same frequency. But remember that as the mirror advances, less room is left for the wave crests both in the incident and reflected waves. The crests that are "squeezed out" each second must cause more crests to leave each second than have arrived.

Show that the two proofs (for particles and waves) can be formulated in the same way if wave number is substituted for momentum and frequency for energy. The only requirement is that frequency should be proportional to energy with the same constant of proportionality as the one connecting wave number and momentum. Note that you do not even have to calculate the energy change and the frequency change due to the reflection. It suffices to show that the same set of equations determines both.

APPENDIX X: SCHROEDINGER'S EQUATION

The Schroedinger equation is the translation of the equation

$$E_{kin} + E_{pot} = E_{tot}$$

where E_{kin} is the kinetic energy (which depends on the momentum of the electron), E_{pot} is the potential energy (which depends on the position of the electron), and E_{tot} is the total energy which is a constant for any definite or "stationary" (that is, time independent) state of the electron. Restrict this consideration to one-dimensional problems so that the wave function depends only on **x**.

1.) Introduce the idea of a "linear operator," **O**, which turns a wave function Ψ into a new function $O\Psi$. More specifically, the operator is linear if for any two wave functions

$$O(\Psi_1 + \Psi_2) = O\Psi_1 + O\Psi_2$$

which are the only ones to be used. Important examples are $x\Psi = x \cdot \Psi$ (read the operator **x** acting on Ψ is **x** times Ψ) and

$$p\Psi = \frac{h}{2\pi i} \frac{d\Psi}{dx}$$

Assume

$$\Psi = e^{2\pi ikx} = \cos 2\pi kx + i \sin 2\pi kx, \text{ where}$$

$(i)^2 = -1$, and $e^{2\pi ikx}$ is defined by the expression given above.

Show that this last relation gives the same result as de Broglie's proposal connecting wave number values with momentum values. (We shall return to this discussion in Appendix XI.)

2.) Show that in classical physics,

$$E_{kin} = \frac{p^2}{2m}$$

where m is the mass of the electron.

3.) Translate

$$\frac{p^2}{2m}$$

into wave-language by the replacement of p^2 by using the operator

$$\frac{h}{2\pi i}$$

times differentiation with respect to x twice in succession. Also retain $1/2m$ as a factor.

4.) Translate E_{pot} into wave-language by replacing E_{pot} (which is a function of x) by the command to multiply E_{pot} and Ψ.

5.) Translate E_{tot} into wave language by multiplying E_{tot} and Ψ which is a simple constant factor times Ψ.

6.) Derive the simplest Schroedinger equation:

$$-\frac{h^2}{4\pi^2}\frac{d^2\Psi}{dx^2} + E_{pot} \cdot \Psi = E_{tot} \cdot \Psi$$

where $\frac{d^2\Psi}{dx^2}$ is the command: differentiate, then differentiate again.

7.) Show that if the function E_{pot} approaches constant values where x becomes very large or −x becomes very large and has smaller values somewhere in between, then Schroedinger's equation has, in general, no solutions that remain finite for arbitrary energy values E_{tot} smaller than the limiting values of E_{pot} at large x and large −x.

This means that in general *bound* states for the electron (where, according to classical ideas, the electron does not have enough energy to escape toward large x or large −x) do not exist; indeed, a state where Ψ grows beyond all limits (and the probability of finding the electron becomes very large precisely where the electron should not be found at all) makes no sense.

8.) Show, however, that for specific sharply defined values of E_{tot}, the Schroedinger equation does give satisfactory bound solutions, that is, solutions where Ψ approaches 0, when x or −x becomes very large. These are the solutions where the wave fits into the region where the potential energy has a minimum. The general proofs of these statements are a little involved. Specific examples can be more easily found by anyone who knows calculus. (Suggestion: choose any reasonable Ψ and E_{tot}; from these calculate E_{pot}.)

So far only the time independent Schroedinger equation has been considered. In a time dependent equation, replace $E_{tot} \cdot \Psi$ by

$$-\frac{h}{2\pi i}\frac{\delta\Psi}{\delta t} \quad \text{(and also replace } \frac{d^2\Psi}{dx^2} \text{ by } \frac{\delta^2\Psi}{\delta x^2}\text{)}$$

where the new kind of differentiation (using small rounded delta symbols) means that one changes t *or* x while leaving the other variable constant.

1.) Show that the replacement of E_{tot} by an operator corresponds to Planck's relation ($E = h\nu$)..

2.) Discuss how the time-dependent Schroedinger equation can be used to evaluate the future behavior of the Ψ function in time, that is how the knowledge of the Ψ function at one moment leads to the knowledge of what it will be in the next moment, and, by repetition of the procedure to the knowledge of what it will be at any later time.

APPENDIX XI: HEISENBERG'S UNCERTAINTY PRINCIPLE

Heisenberg's uncertainty principle, $\Delta x \cdot \Delta p_x$ is equal to or greater than

$$\frac{h}{2\pi}$$

(with Δx the uncertainty of x coordinate and Δp_x the uncertainty in the x-component of the momentum) can be understood in the following steps. (The dot following Δx means multiplication.)

1.) Construct a wave function by adding sinusoidal waves all of which reinforce each other at a given point.

2.) Assume that we have used (with considerable amplitude) waves $e^{2\pi ikx}$, where k covers the range Δk (where k is the wave-number, that is, the wave maxima encountered per unit of distance); show that the waves tend to cancel when one has moved away from the point of mutual reinforcement by the distance $1/2\pi\Delta k$. Hence $\Delta x \cdot \Delta k$ cannot become less than $1/2\pi$.

3.) Consider the connection between wave number and momentum explained below and also implicit in Appendix X (concerning Schroedinger's equation), and derive the relation between Δx and $\Delta p_x = h\Delta k$.

4.) Note that the wave functions can be written with arbitrarily high values of Δx and Δp_x. Heisenberg makes statements only about *minimum* values of the uncertainties in the product.

All this can be seen in greater detail in the two extreme cases (considering only one dimension, x, and one momentum component, p_x). Two states are to be considered:

a. Position known accurately; momentum unknown;
b. Momentum known accurately; position unknown.

These cases are simple but need peculiar arguments. The following somewhat difficult considerations are recommended.

1.) Think of a wave function Ψ which is zero everywhere except near the position where the electron actually is to be found, for instance, at $x=0$. The wave consists of a single maximum or hump. Contract the wave function more and more, so that the electron should be more and more accurately found precisely at $x = 0$. At the same time make the hump higher and higher so that when one plots the wave function squared, $|\Psi|^2$, as a function of x, the area under the curve remains unity. This means that the probability of finding the electron anywhere is 1 (or, in other words, the electron is *somewhere* with certainty). The function so defined is called the Dirac delta function.

Now show (and this is not easy for the beginner) that if one adds cosine waves, $\cos 2\pi \mathbf{kx}$, where **k** is the wave number and where all waves have their maximum value = 1 at $x = 0$, one will get a function similar to the Dirac delta function. (Only if one adds, or integrates, over all **k**-values, the height of the hump becomes infinite much too fast.) According to de Broglie, **hk** is the momentum. So for the delta function, all momentum values are obtained.

2.) If the momentum-value is known, the wave function has the form

$$e^{2\pi ikx} = \cos 2\pi kx + i \sin 2\pi kx,$$

where **i** is the square root of -1. Note that $\cos 2\pi kx$ or $\sin 2\pi kx$ have no reason to represent motion toward increasing or decreasing x-values. But

$$e^{2\pi ikx} \text{ and } e^{-2\pi ikx}$$

(which is equal to $\cos -2\pi ikx - i \sin 2\pi kx$) can be better identified with positive and negative momentum (corresponding to **k** and **-k**) values respectively. What matters here is that

$$|e^{2\pi ikx}|^2 = 1,$$

that is, the electron is found at each value of x with the same probability while the value of the momentum **hk**, together with the direction of the momentum is sharply defined. Note also that the sum (or rather integral) of all probabilities will again be infinite so that the wave functions presented here are not quite properly defined.

3.) Now consider a real case

$$\Psi = ae^{2\pi ikx}e^{-(x/d)^2}$$

Here "a" is a constant which must be so adjusted that the integral (or effective summation) over all $|\Psi|^2$ -values should give the total probability of 1. Represent Ψ as a sum (or integral) over sine and cosine functions. This will give the range of **k**-values and the range of momentum values. Compare this with the range of **x**-values that one obtains from the factor

$$e^{-(x/d)^2}$$

Show that an appropriate average over these ranges or uncertainties gives the Heisenberg relation

$$\Delta x \cdot \Delta p = \frac{h}{2\pi}$$

All these procedures are relatively easy after about five years practice. For the unmathematical reader, they can hardly give more than an impression of how the uncertainty relation is connected with wave functions.

APPENDIX XII: PROVING FERMAT'S THEOREM REGARDING PRIMES AS THE SUM OF SQUARES

The easy part:

1.) If a prime $p = 4n - 1$ equals $a^2 + b^2$ (a and b are integers), then one of the two numbers a and b must be even; the other is odd. Let a be even ($a = 2n$) and b odd ($b = 2n + 1$);

2.) a^2 is divisible by 4;

3.) b^2 is four times an integer *plus* 1;

4.) $a^2 + b^2$ is four times an integer *plus* 1.

The difficult part:

1.) Prove that any number can be written in one and only one way as a product of primes. (The proof is well known but difficult. The easiest proof makes use of the "Euclidean algorithm" which is a straightforward method to find the greatest common denominator of two integers.)

2.) Introduce the imaginary unit **i** which is the square root of −1.
3.) Introduce complex integers which have the form **a + ib** where **a** and **b** are integers.
4.) Introduce complex primes which can be divided only by 1, −1, **i**, themselves or products of these with the complex prime.
5.) Prove that any complex integer can be written (apart from factor −1 and **i**) only in one way as a product of complex primes. (The proof is similar to that for ordinary or real primes.)
6.) Note that a real prime is also a complex prime unless it can be written as $p = a^2 + b^2$. In the latter case, $p = (a + ib)(a - ib)$ and is, therefore, not a complex prime. (Now one can see why the question whether or not **p** can be written as a sum of squares is interesting.)
7.) Using the theorems of unique prime decomposition, show that a real prime cannot be written in more than one way as a sum of squares.
8.) Introduce the notion "congruence modulo p" (with p equal to a real prime) which is written $a \equiv b \bmod p$ where a and **b** are integers; $a \equiv b \bmod p$ means that $a - b$ is divisible by p. We can now play with p numbers: 0, 1, 2, . . . , p − 1.
9.) Prove "Fermat's little theorem:" $x^{p-1} \equiv 1 \bmod p$. (**x** is an integer.) This is only moderately difficult.
10.) Show that the congruence $x^m \equiv 1 \bmod p$ cannot have more than **m** different solutions.
11.) Note that $x^{p-1} \equiv 1 \bmod p$ has precisely $p - 1$ solutions, namely 1, 2, . . . , p − 1. . .
12.) For $p = 4n + 1$, consider all numbers $x^n \equiv y \bmod p$. Note that $y^4 \equiv 1 \bmod p$.
13.) $y^4 \equiv 1 \bmod p$ has the solutions $y = 1$ and $y = p-1$. Using all numbers which satisfy $x^n \equiv 1 \bmod p$ and $x^n \equiv p-1 \bmod p$, one gets at most **2n** values for **x**. Show that **y** values must exist which satisfy $y^4 \equiv 1 \bmod p$ where **y** is not 1 or −1.
14.) Assume $p = 4n + 1$ (**n** is real) is a complex prime.
15.) Apply the theorems about congruences for complex integers with $a = 0, 1, 2, \ldots, p-1$ and $b = 0, 1, 2, \ldots, p-1$.
16.) Solve the equation $y^4 \equiv 1 \bmod p$ for complex primes. The numbers 1 and $p - 1$ and, according to step 13, at least one more real y-value are solutions. Two more solutions are **i** and $i(p - 1)$. Therefore, $y^4 \equiv 1 \bmod p$ has at least 5 solutions which contradicts 10.
17.) Therefore, assumption 14.) is wrong.
18.) If $p = 4n + 1$, a real prime is not a complex prime. Prove that it can be written as $p = (a + ib)(a - ib) = a^2 + b^2$.

The reader should observe the contrast between the complexity of the proof and the relative simplicity of the result. In establishing the proof, s/he has learned a few things about systems of prime numbers and operations with prime numbers. In other words, if the reader has been successful, his or her understanding of prime numbers has improved.

Index